Advance Praise for
Power from the Sun

Dan has done it again with his Second Edition of *Power from the Sun*.
It's a thorough and comprehensive compendium from the basics to the complex
aspects of solar and the imperative to continue our conversion to a renewable
energy powered economy if our species is to survive into the 22nd Century.
Dan's prolific writing over the decades has been a gift to all of us.
A brilliant treatise on solar and a perfect companion to
the *Real Goods Solar Living Sourcebook!*
—John Schaeffer, Founder & Owner of Real Goods & author of
the *Solar Living Sourcebook*

Dan Chiras has been a solar pioneer for decades. The depth and
breadth of his understanding of solar technologies, policies, and specification
and installation best practices is unmatched. This book is a comprehensive
guide to all things solar that everyone interested in the space should
read, especially since solar is playing such a vital role in the
transition to a sustainable economy.
—Sara Gutterman, CEO, Green Builder Media

Dan Chiras clearly communicates theory and practice of a resource
critically important to integrate into our national energy mix.
—Jean Ponzi, Green Resources Manager,
EarthWays Center of Missouri Botanical Garden

Dan Chiras makes it as easy as possible for you to effect
your own transition away from fossil fuel dependence. When you
need practical advice from a warm, smart and informed
human being, Dan Chiras is the one to turn to.
— Bruce King, PE Director, Ecological Building Network, and author,
Buildings of Earth and Straw and *Making Better Concrete*

Dan Chiras strikes again! His books — which occupy a whole shelf in my bookcase — have now enabled thousands of energy activists to convert sun, wind, and water into self-reliance.

— David Wann, coauthor, *Affluenza* and *Superbia!*, and author, *Simple Prosperity* and *The New Normal*

Dan Chiras is as reliable as a Swiss watch — once again he's created a text that's as accessible as it is informative.

— Ann Edminster, author, *Energy Free: Homes for a Small Planet*

Power
from the
Sun

A Practical Guide to
Solar Electricity

REVISED 2ND EDITION

DAN CHIRAS

new society
PUBLISHERS

Cover design by Diane McIntosh. Cover image © iStock

Printed in Canada by Friesens.
First printing October, 2016.

Inquiries regarding requests to reprint all or part of *Power from the Sun*
should be addressed to New Society Publishers at the address below.

To order directly from the publishers, please call toll-free (North America)
1-800-567-6772, or order online at www.newsociety.com

Any other inquiries can be directed by mail to:

New Society Publishers
P.O. Box 189, Gabriola Island, BC V0R 1X0, Canada
(250) 247-9737

LIBRARY AND ARCHIVES CANADA CATALOGUING IN PUBLICATION

Chiras, Daniel D., author
Power from the sun : a practical guide to solar electricity
/ Dan Chiras. — Revised 2nd edition.

Includes bibliographical references and index.
Issued in print and electronic formats.
ISBN 978-0-86571-829-6 (paperback). — ISBN 978-1-55092-624-8 (ebook)

1. Photovoltaic power systems.   2. Photovoltaic power generation.
3. Solar Houses,   4. Dwellings—Power supply.   I. Title.

TK1087.C45 2016   621.31'244   C2016-905827-1
                                             C2016-905828-X

Funded by the   Financé par le
Government   gouvernement
of Canada   du Canada

New Society Publishers' mission is to publish books that contribute in fundamental ways
to building an ecologically sustainable and just society, and to do so with the least possible
impact on the environment, in a manner that models this vision. www.newsociety.com

FSC
www.fsc.org
MIX
Paper from
responsible sources
FSC® C016245

Certified
B
Corporation

new society
PUBLISHERS

Contents

CHAPTER 1

An Introduction to Solar Electricity

Like many other people throughout the world, I have been dreaming of a future powered by renewable energy for many years. Fortunately, that dream is becoming a reality. Today, solar and wind energy have become major contributors to global energy production. In the United States alone, solar, wind (mostly large-scale wind), biomass, and geothermal now produce about 7% of the nation's electricity. When you add hydroelectricity to the mix, renewable energy now supplies over 13% of America's electricity. That's a recent change, occurring within the last few years. Although we have a ways to go, that's pretty amazing.

What is even more amazing is that some states now generate about 25% of their electricity from renewables. Kansas and Iowa, for example, both do, mostly with wind power. Colorado generates 18% of its electricity from renewables, but its goal is 30% by 2020. California is aiming for 33% by 2020. While impressive, get this: Germany now produces over 60% of its electricity from renewables.

Although there are many reasons why renewables have gone wild in recent years, one of them has been cost. For years, renewable energy advocates have dreamed of a day when we could say that renewables cost the same or less than conventional energy resources. At least for solar electricity, those days are here. Solar electric systems, even without subsidies, often produce electricity at or below the cost of power from utilities. They have reached price parity, and the future is looking quite bright.

This book focuses on an important element of the renewable energy dream: solar electricity. Solar electric systems are also known as *photovoltaic systems* or *PV systems*, for short. In this book, I will focus primarily on residential-scale PV systems, that is, systems suitable for homes and small businesses. These systems generally fall in the range of 1,000 watts for very small, energy-efficient cabins or cottages to 5,000 to 15,000 watts for typical suburban homes. All-electric homes could require even larger systems, on the order of 25,000 watts. What does all this mean?

Understanding Rated Power

One of the most frequent questions solar installers are asked is "What size system do I need?" It's probably the question that's burning in your mind right now.

The answer is almost always the same: it depends. More specifically, it depends on how much electricity you and your family use. Once an installer knows how much electricity you consume, he or she can size a system.

The size of a solar electric system is given in its *rated power*, also known as *rated capacity*. Rated power of a system is the output of the modules (PV panels) times the number of modules. If your system contains twenty 250-watt modules, its rated capacity is 5,000 watts. Because *kilo* means 1,000, a 5,000-watt system is also called a 5-kilowatt, or 5-kW system. But what does it mean when I say a module's rated power is 250 watts?

Rated power is the instantaneous output of a solar module measured in watts under standard test conditions (STC). Watts is a measure of power. Another way of thinking about it is as the rate of the flow of energy. The higher the wattage, the greater the flow.

Watts is also the measurement used to rate technologies that produce electricity, such as a solar module or a power plant. For example, most solar modules installed these days are rated at 250 to 300 watts. Power plants are rated in millions of watts, or megawatts: a typical coal-fired power plant produces 500 to 1,000 megawatts of power.

Most readers are probably more familiar with the terms watt and wattage when they are used to rate power consumption of a device, such as a light bulb or a hair dryer. You may have a 12-watt compact fluorescent light bulb, for instance, or an 1,800-watt hair dryer. Your microwave may be 1,200 watts. In such cases, wattage indicates power consumption. The higher the wattage, the greater the consumption.

Where Does Your Electricity Come From?

Did you ever wonder where your electricity comes from? According to the US Energy Information Administration (EIA), in 2014, electricity sold by utilities came from the following sources:

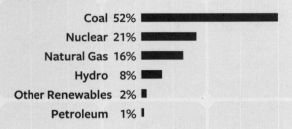

Coal	52%
Nuclear	21%
Natural Gas	16%
Hydro	8%
Other Renewables	2%
Petroleum	1%

As in many cases, national averages can be deceiving. Your utility may rely almost entirely on coal, nuclear, or hydropower. In the Midwest, for instance, it is not uncommon for 75% of one's electrical energy to come from coal. In the Pacific Northwest, the vast majority of the electricity comes from hydropower. You may be able to find out where your electricity comes from by calling your local utility and talking with their public information specialists.

As just noted, the rated output of a PV module is determined under standard test conditions—a set of conditions that all manufacturers use to rate their modules. To test a module, workers set it up in a room that is maintained at 77°F (25°C). A light is flashed on the module at an intensity of 1,000 watts per square meter. That is equivalent to full sun on a cloudless day in most parts of the United States. The light is arranged so light rays strike the module at a 90° angle, that is, perpendicularly. (Light rays striking the module perpendicular to its surface result in the greatest absorption of the light.)

Rating modules under standard test conditions is useful for comparing one manufacturer's module to another's. A 285-watt module from company A should perform the same as a 285-watt module from company B.

Installers can use this rating to determine the size of a system required to meet your needs. As noted above, most residential solar electric systems fall within the range of 5,000 to 15,000 watts, or 5 to 15 kW. It is important to note, however, that a 5-kW PV system won't produce 5,000 watts of electricity all the time the Sun's shining on it. This rating is based on standard test conditions, which are rarely duplicated in the real world. So, this rating system leaves a bit to be desired.

When mounted outdoors, PV modules typically operate at higher temperatures than those subjected to standard test conditions. That's because infrared radiation in sunlight striking PV modules causes them to heat up. To achieve a cell temperature of 77°F (25°C) under full sun, like that measured in the laboratory, the air temperature must be quite low—about 23°F to 32°F (0°C to −5°C), not a typical temperature for PV modules in most locations most of the year. This is important to know because higher temperatures decrease the output of a solar module. That's the reason why it is so important to mount a solar system so that the modules stay as cool as possible. It also explains why solar modules on a ground-mounted rack typically perform better than roof-mounted systems. Roofs are hotter.

To better simulate real-world conditions, the solar industry has developed an alternative rating system. They call it *PTC*, which stands for *PV-USA Test Conditions* (which stands for *PV for Utility-Scale Applications Test Conditions*).

PTC were developed at the PV-USA test site in Davis, California, and more closely approximate real-life operating conditions. In this rating system, the ambient temperature is held at 68°F (20°C). The modules face the Sun and output is measured when the Sun's irradiance reaches 1,000 watts per square meter. The conditions take into account higher module temperatures and the cooling effect of wind. Wind speed in the test is 2.24 miles per hour (or 1 meter per second) 10 meters above the surface of the ground. (That is, wind speed is monitored at 10 meters, or

33 feet.) Although the cell temperature varies with different modules, it is typically about 113°F (45°C) under PTC. Because power output (in watts) decreases with rising temperature, PTC ratings are roughly 10% lower than STC ratings.

When shopping for PV modules, I used to suggest that readers ignore the name-plate rating the manufacturers derived under STC and look up the PTC ratings. Some manufacturers publish these data on their spec sheets; others don't. As a rule, though, it's not that important. Solar installers will estimate the output of an array taking into account local conditions. If you must, you can find PTC ratings on the California Energy Commission's website: consumerenergycenter.org.

An Overview of Solar Systems

As their name implies, solar electric systems convert sunlight energy into electricity. This conversion takes place in the solar modules, more commonly referred to as *solar panels*. (The term *modules* is the correct one; that's what they are called in the National Electric Code, but few people use this term.)

A solar module consists of numerous *solar cells*. Most solar modules contain 60 solar cells, although manufacturers are also producing larger modules that contain 72 cells each. Solar cells are made from one of the most abundant chemical

FIGURE 1.1. These solar modules being installed by several students at The Evergreen Institute consist of numerous square solar cells. Each solar cell has a voltage of around 0.6 volts. The cells are wired in series to produce higher voltage, which helps move the electrons from cell to cell. Modules are also wired in series to increase voltage. Credit: Dan Chiras.

substances on Earth, silicon dioxide. (Silicon dioxide makes up 26% of the Earth's crust.) Silicon dioxide is found in quartz and from a type of sand that contains quartz particles. Silicon is extracted from silicon dioxide.

Solar cells are wired in series inside a module (see sidebar). The modules are encased in glass (in the front) and usually plastic (in the back). The plastic and glass layers protect the solar cells from the elements, especially moisture. Most modules these days have aluminum frames. Because the anodized (silver) aluminum frames are striking, many manufacturers produce modules with black aluminum frames. The aluminum rails and mounting hardware used to attach them to the roof are also black, so the array blends in well, and homeowners associations and neighbors are happy. Some manufacturers are now producing frameless modules to make them even more aesthetically appealing. One manufacturer's modules passed an unprecedented Class 4 hail testing (that is, they withstood 2-inch hail striking it at 75 mph), making them the most impact-resistant module in the industry. Frameless modules have been designed to make installation quick, easy, and less expensive.

Understanding Electricity: Series Wiring

Although few people know what series wiring is, we use it all the time. Every time you put batteries in a flashlight, for instance, you place the positive end of one battery against the negative end of the next one. Series wiring increases the voltage. Voltage is a force that moves electrons (it is considered an *electromotive* force). The higher the voltage, the more force. When placing two 1.5-volt batteries in series in a flashlight, you increase the voltage to 3. Place four 1.5-volt batteries in series and the voltage is 6.

In solar modules, each solar cell has a positive and a negative lead (wire). To wire in series, the positive lead of one module is soldered to the negative of the next, and so on and so on. Most solar cells in use today are square and measure 5 × 5 inches (125 × 125 mm) or 6 × 6 inches (156 × 156 mm) and have a voltage of about 0.6 volts. When manufactures wire (actually, solder) 60 solar cells together in series in a module, the voltage increases to 36. Wiring several solar modules together in series, in turn, increases the voltage of an array. Ten 36-volt modules wired in series results in an array voltage of around 360.

A group of modules wired in series is known as a *series string*. Most residential solar arrays consist of series strings of 10 to 12 modules. String sizes in commercial PV array are larger.

FIGURE 1.2. Solar electric systems produce direct current (DC) electricity just like batteries. As shown here, in DC circuits the electrons flow in one direction. The energy the electrons carry is used to power loads like light bulbs, heaters, and electronics. Credit: Forrest Chiras.

Two or more modules are typically mounted on a rack and wired together. Together, the rack and solar modules are referred to as an *array*. Ground-mounted arrays are typically anchored to the earth by a steel-reinforced concrete foundation. Together, the rack, modules, and foundation are referred to as an *array*.

Solar modules produce direct current (DC) electricity. That's one of two types of electricity in use today. It consists of a flow of tiny subatomic particles called *electrons*. They flow through conductors, usually copper wires. In direct current electricity, electrons flow in one direction, as illustrated in Figure 1.2. In this case, electrons flow out of the battery through the light bulb, where they give off their energy, creating light and heat. The de-energized electrons then flow back into the battery.

Electricity generated by a PV array flows via wires to yet another component of the system, the *inverter*. This remarkable device converts DC electricity produced by solar cells into alternating current (AC) electricity. AC is the type of electricity used in homes and businesses throughout most of the world.

Applications

Once a curiosity, solar electric systems are becoming commonplace throughout the world. While many solar electric systems are being installed to provide electricity to homes, they are also being installed

FIGURE 1.3. This solar array was installed at the Owensville Middle School in Owensville, Missouri, by the author and his business partner, Tom Bruns, and numerous hard-working students. The array is small for a school of this size, which uses almost 1,000 kilowatt-hours per day; the array only provides about 12 days' worth of electricity to the school. The array is also used for science and math education. Credit: Dan Chiras.

on schools, small businesses, office buildings, and even skyscrapers, like 4 Times Square in New York City, home of NASDAQ. Figure 1.3 shows a photo of a solar array my company installed at Owensville Middle School in Owensville, Missouri, with the help of several eager middle-schoolers. Many large corporations such as Microsoft, Toyota, and Google have also installed large solar electric systems. More and more electric utilities are installing large PV arrays to supplement conventional sources.

Even colleges are getting in on the act. Colorado College, where I once taught courses on renewable energy, installed a large solar system on one of its dormitories. And over the years, several airports have installed large solar arrays to help meet their needs. At Sacramento's airport, for instance, PVs were built to create parking structures that shade vehicles during the day while also generating electricity. Denver International Airport installed a 2,000 kilowatt solar electric system in 2008, and it has greatly expanded its system since then.

On a smaller scale, ranchers often install solar electric systems to power electric fences to contain cattle and other livestock. Many have installed small solar systems to pump water for their stock, saving huge amounts of money on installation. We power our farm and my educational center, The Evergreen Institute, on solar electricity. I installed a solar pond aerator to lengthen the lifespan of our pond and keep it open during the winter (Figure 1.4).

Solar electricity has proven extremely useful on boats. Sailboats, for example, are often equipped with small PV systems—usually a few hundred watts—to power lights, fans, radio communications, GPS systems, and small refrigerators. Many recreational vehicles (RVs) are equipped with small PV systems that are used to power microwaves, TVs,

FIGURE 1.4. (*above*) Aerating my half-acre farm pond reduces bottom sludge and helps create a healthier environment for the fish. (*below*) To prevent the pond from icing over completely in the winter (so our ducks can swim), I designed and installed a solar pond aerator in conjunction with a St. Louis company, Outdoor Water Solutions. They are currently selling a design like this online. Credit: Dan Chiras.

FIGURE 1.5. This electronic road sign is powered by a small solar array. Credit: Dan Chiras.

FIGURE 1.6. These bikes in downtown Denver, Colorado, can be rented by the hour. The payment system (*left*) is powered by solar electricity. Credit: Dan Chiras.

and satellite receivers. The electricity these systems produce reduces the run time for gas-powered generators that can disturb fellow "campers."

Many bus stops, parking lots, and roadways are now illuminated by superefficient lights powered by solar electricity, as are portable information signs used at highway construction sites (Figure 1.5). As electric cars become more and more popular, many homeowners have set up solar arrays to charge their vehicles. (Others, like me, just use the solar supplied by the arrays that power our homes or businesses.)

Numerous police departments now haul solar-powered radar units to neighborhoods to discourage speeding. These units display a car's speed and warn drivers if they're exceeding the speed limit.

The trash compactors at the Kennedy Space Center are powered by solar electricity. In downtown Colorado Springs, sidewalk parking meter pay stations are powered by solar. In Denver, credit card payment centers for city bikes that people can rent for the day are powered by solar (Figure 1.6). At the convention center in Topeka, Kansas, you can rent a bike, if you need some temporary transportation. Each bike is equipped with a small solar pay station.

Backpackers and river runners can take small solar chargers or larger lightweight fold-up solar modules with them on their ventures into the wild to power electronics. Military personnel have access to similar products, and there are numerous portable devices available for charging cell phones and tablets in our daily lives. Some are even sewn into backpacks, sports bags, or briefcases. There are also a number of cart- and trailer-mounted PV systems for emergency power after disasters. (To learn more, read Jeff Yago's article in *Home Power*, Issue 168.)

Solar electric systems are well suited for remote sites where it is often cost prohibitive to run power lines, such as cottages, cabins, and superefficient off-grid homes. In France, the government paid to install solar electric systems and wind turbines on farms at the base of the Pyrenees Mountains, rather than running electric lines to these distant operations.

Solar electricity is also used to power remote monitoring stations that collect data on rainfall, temperature, and seismic activity. PVs allow scientists to transmit data back to their labs from remote sites, like the tropical rain forests of South

Understanding Electricity: Power vs. Energy

If you are going to install a PV system on your home or become a professional installer, it is important to understand two very important terms: power and energy. In the electrical world, power is a measure of the flow of energy, in our case electrical energy. Power is measured in watts or kilowatts. (Remember: 1,000 watts is 1 kilowatt.) Power is an instantaneous measure. That is, it is like the speed of a car. If you glance down at your speedometer, and it reads 60 miles per hour (96 kph) you have just determined the rate or speed at which your car is traveling at that instant. A deer may dart across the road, causing you to step on the brakes, and the rate of speed will decrease. Shaken by the near miss, you might drive at 50 miles per hour (80 kph) for a while.

Also noted in the previous sidebar, engineers and scientists rate electrical loads (devices that consume electricity) such as light bulbs and electric motors in watts or kilowatts. For example, a light bulb might be rated at 12 watts. An electric motor might be rated at 1,000 watts. An element in an electric water heater might be rated at 4,500 watts.

Scientists and engineers also use watts to rate electric-generating technologies, such as solar electric systems. A small PV system for an energy-efficient home, for instance, might be rated at 5,000 watts. A larger, less-efficient home might require a 15,000-watts or 15-kW system.

Energy, in contrast, is power consumption or production over time. Put another way, it is a quantity. Something you can measure. For the mathematically inclined, it is the product of rate × time. For example, suppose a CFL light bulb consumes 12 watts for one hour. The amount of electricity consumed is expressed in watt-hours. To determine watt-hours, multiply watts by time in hours. In this example, a 12-watt light bulb that operates for 1 hour consumes 12 watt-hours of electricity. If it operates for 10 hours, it uses 120 watt-hours of energy or 0.12 kilowatt-hours (kWh) of energy.

Both watt-hours and kilowatt-hours are measures of energy use. Utilities charge us for the energy we consume—the amount of power we consume—each month. Your utility probably charges somewhere between 8 and 30 cents per kWh of electricity you use.

While we consume energy in our homes, PV systems produce energy. A solar electric system producing electricity at a rate of 1,000 watts for a period of one hour produces 1,000 watt-hours or 1 kWh (kilowatt-hours) of energy. If it produces 1,000 watts for four hours, it has produced 4 kWh. If it produces 500 watts for two more hours, the daily total is 5 kWh.

America. Stream flow monitors on many rivers and streams throughout the world rely on solar-powered transmitters to beam data to solar-powered satellites. In the United States, this data is then beamed back to Earth to the US Geological Survey, where it is processed and disseminated.

Solar electric modules often power lights on buoys, vital for nighttime navigation on large rivers like the Saint Lawrence Seaway. Railroad signals and aircraft warning beacons are also often solar-powered.

PV modules are used to boost radio, television, and telephone signals. Signals from these sources are often transmitted over long distances. For successful transmission, however, they must be periodically amplified at relay towers. The towers are often situated in inaccessible locations, far from power lines. Because they are reliable and require little, if any, maintenance, PV systems are ideal for such applications. They make it possible for us to communicate across long distances. Next time you make a long-distance telephone call from a phone on a landline, rest assured solar energy is making it possible.

While PV systems are becoming very popular in more developed countries, they're also widely used in the developing world. They are, for instance, being installed in remote villages in less developed countries to power lights and computers and the refrigerators and freezers used to store vaccines and other medicine. They're also used to power water pumps.

The ultimate in remote and mobile applications, however, has to be the satellite. Virtually all military and telecommunications satellites are powered by solar electricity, as is the International Space Station.

World Solar Energy Resources

Solar electricity is rapidly growing in popularity, which is fortunate because global supplies of fossil-fuel resources like coal, natural gas, and oil are on the decline. As they decline, and as concern over global climate change increases, solar electric systems along with small and large wind systems and other forms of renewable energy farms will become a major source of electricity throughout the world. It's inevitable. But is there enough solar energy to meet our needs?

Although solar energy is unevenly distributed over the Earth's surface, significant resources are found on every continent. "Solar energy's potential is off the chart," write energy experts Ken Zweibel, James Mason, and Vasilis Fthenakis in a December 2007 article, "A Solar Grand Plan," published in *Scientific American*. Less than one billionth of the Sun's energy strikes the Earth, but, as they point out, the solar energy striking the Earth in a 40-minute period is equal to all the energy human society consumes in a year. That is, 40 minutes of solar energy is equivalent to all the coal, oil, natural gas, oil shale, tars sands, hydropower, and wood we consume in an entire year. What is more, we'd only need to capture about 0.01% of the solar energy striking the Earth to meet *all* of our energy demands. Solar electric systems mounted on our homes and businesses or in giant commercial solar arrays could tap into the Sun's generous supply of energy, providing us with an abundance of electricity.

Could solar electric provide 100% of the United States' or the world's electrical energy needs?

Yes, it could.

Will it?

Probably not.

In fact, no one is planning on a 100% solar future. Rather, most renewable energy experts envision a system that consists of a mix of renewable energy technologies such as solar hot water, solar electricity, passive solar, wind, hydro, geothermal, and biomass. These renewable energy resources, combined with radical improvements in energy efficiency, will become the mainstay of the world's energy production.

In a sustainable global energy system, PV systems could play a significant role. They could produce enormous amounts of electricity for homes, businesses, farms, ranches, schools, and factories. Another solar technology, large-scale solar thermal electric systems, could supplement PV systems and play a role in meeting our needs. Solar thermal electric systems concentrate sunlight energy to generate heat that's used to boil water. Steam generated from this process is used to spin a turbine connected to generator that makes electricity (Figure 1.7). Some of the newest solar thermal electric systems even store hot water so electricity can be generated on cloudy days or in the evening.

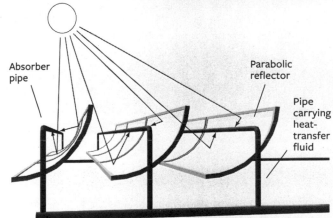

FIGURE 1.7. (*top*) Solar thermal electric systems like the one shown here reflect sunlight off of highly polished parabolic reflectors onto pipes (*bottom*) that contain heat-transfer fluid that absorbs energy from the intense, focused light. The heat is transferred via a heat exchanger to a large, well-insulated tank where it boils, producing steam. Steam turns a turbine that generates electricity much the same as in a conventional power plant. Although solar thermal electric systems can produce electricity at rates competitive with conventional power plants, they require a lot more maintenance than traditional solar electric systems. Photo Credit: Sandia National Laboratory. Art Credit: Anil Rao.

It is likely that large-scale wind farms will provide more electricity to power our future than PVs. (The world currently produces three times as much electricity from wind as it does from solar). Geothermal and biomass resources could contribute as well. *Biomass* is plant matter that can either be burned directly to produce heat or to generate steam that's used to power a turbine that generates make electricity. Biomass can also be converted to liquid or gaseous fuels that can be burned to produce electricity or heat (or power vehicles). Hydropower currently contributes a significant amount of electricity throughout the world, and it will continue to add to the energy mix in our future.

What will happen to conventional fuels such as oil, natural gas, coal, and nuclear energy? Although their role will diminish over time, these fuels will very likely contribute to the energy mix for many years to come. In the future, however, they will very likely take a back seat to solar, wind, and other renewables. They could eventually become pinch hitters to renewable energy.

Despite what many ill-informed critics say, renewable energy is splendidly abundant. What is more, the technologies needed to efficiently capture and convert solar energy to useful forms of energy like heat, light, and electricity are available now and, for the most part, quite affordable. Costs have plummeted in recent years. Take solar electricity, for example: the cost of solar electric modules has fallen from over $75 per watt in 1977 to $0.72 in 2015 (Figure 1.8).

While nuclear and fossil fuels are on the decline, the Sun will enjoy a long future. The Sun, say scientists, will continue to shine for at least five billion more years. To be truthful, however, scientists estimate that the Sun's output will increase by about 10% in one billion years, making planet Earth too hot to sustain life. So, we don't

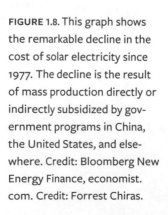

FIGURE 1.8. This graph shows the remarkable decline in the cost of solar electricity since 1977. The decline is the result of mass production directly or indirectly subsidized by government programs in China, the United States, and elsewhere. Credit: Bloomberg New Energy Finance, economist. com. Credit: Forrest Chiras.

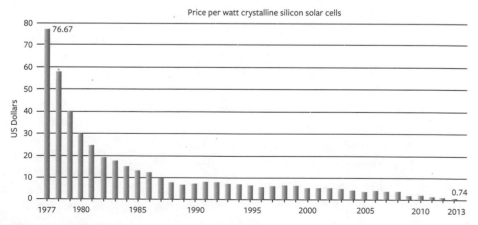

Price per watt crystalline silicon solar cells

US Dollars

76.67

0.74

1977   1980   1985   1990   1995   2000   2005   2010   2013 have five billion years of sunshine, we have less than a billion years before we'll have to check out. But that's a lot more than the 30 to 50 years of oil we have left.

What the Critics Say

Proponents of a solar-powered future view solar energy as an ideal fuel source. It's clean. It's free. It's abundant. And it will be available for a long, long time. Its use could ease many of the world's most pressing problems, such as global climate change, species extinction, and desertification.

Like all fuels, solar energy is not perfect. Critics like to point out that, unlike conventional resources such as coal, the Sun is not available 24 hours a day. Some people don't like the looks of solar electric systems. And, for years, solar electric systems have been pretty pricey, too. Let's take a look at these arguments and respond to the criticisms.

Availability and Variability

Although the Sun shines 24 hours a day and beams down on the Earth at all times, half the planet is always blanketed in darkness. This poses a problem for humankind, especially those of us in the more developed countries, as we consume electricity 24 hours a day, 365 days a year.

Another problem is the daily variability of solar energy. That is, even during daylight hours, clouds can block the sun, sometimes for days on end. At night, PVs produce no energy at all. If solar electric systems are unable to generate electricity 24 hours a day like coal-fired and natural-gas-fired power plants, how can we use them to meet our 24-hour-per-day demand for electricity?

Some homeowners who live off grid (that is, not connected to the electrical grid) solve the problem by using batteries to store the electricity needed to meet nighttime use and the demand on cloudy days. (Remember, however, even on a cloudy day, a solar system will often produce 10% to 20% of its rated output.) I lived off grid for 14 years, and had electricity 24 hours a day, 365 days a year—all supplied by my PV system. I rarely ran out. And when I did, I fired up my backup generator.

One thing all off-gridders learn is that batteries don't store a lot of electricity. In order to live this way, you need to be frugal. It is doubtful that batteries could serve as a backup for modern society. We use—and waste—way too much electricity. We'd need gargantuan, costly battery banks to store energy for nighttime use. Although many strides have been made to store solar electricity, the scale at which we'd have to store it suggest we need an alternative. What is that?

The answer is *coupling*.

Put another way, solar's less-than-24-hour-per-day availability can also be offset by coupling solar electric systems with other renewable energy sources, for example, wind-electric systems, or hydroelectric systems. Wind systems, for instance, generate electricity day and night—so long as the winds blow. In some areas, winds are fairly constant throughout the year. Wind farms in these areas could make up for solar's nighttime absence. Hydroelectric systems tap the energy of flowing water in streams or rivers. These systems range in size from tiny to massive. Microhydro and macrohydro and everything in between can also be used to generate electricity to supplement solar systems, compensating for the Sun's "shortcomings," if you want to call it that.

Solar and wind are especially good partners, as I have found out at my home and educational center in Missouri. Figure 1.9 shows data on the solar and wind resources at my place. As illustrated, the Sun shines quite a lot in the spring, summer, and early fall, but less so during the winter. During winter, however, the winds blow more often and more forcefully. My wind turbine easily makes up for the reduced output of my PV systems during the less-sunny periods of the year, ensuring a reliable, year-round supply of electricity. The same coupling of solar and wind could provide a reliable source of electricity to much of the world.

FIGURE 1.9. This graph of solar and wind energy resources at The Evergreen Institute illustrates how wind and solar energy complement each other. As shown here, sunlight is abundant in the late spring, summer, and early fall in our east-central Missouri location. Wind is most abundant in the late fall, winter, and early spring. This is a common scenario for many North American locations. Credit: Anil Rao.

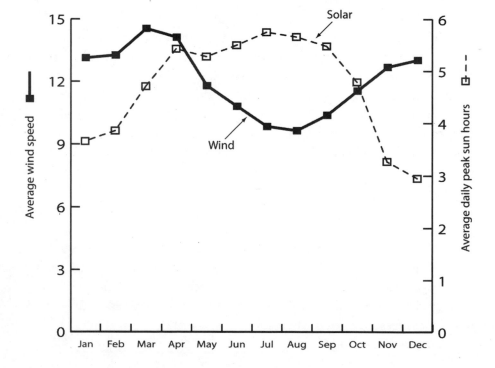

Renewable energy coupling is much more efficient than all the fanciful schemes being proposed to store electricity from solar and wind systems. For example, some individuals have proposed using excess solar or wind electricity to power massive air compressors that pump compressed air into abandoned underground mines. When electricity is needed, the compressed air would be released through a turbine, not unlike those found in conventional power plants. The blades of the turbine would be attached to a shaft that is attached to a generator that produces electricity. While an interesting idea, this plan requires installation of massive solar and wind farms in mining country. Sites may not be suitable for massive renewable energy farms. Mines would also have to be pretty airtight to ensure efficient storage.

Solar electricity could also possibly be used to generate hydrogen gas from water. Hydrogen gas is created when electricity is run through water. This process, known as *electrolysis*, splits the water into its components, hydrogen and oxygen, both gases. Hydrogen can be stored in tanks and later burned to produce hot air or to heat water to produce steam. Hot air and steam can be used to spin a turbine attached to a generator. Hydrogen and oxygen can also be fed into fuel cells, devices that produce electricity.

While schemes such as these could work, they would be nowhere near as efficient as direct renewable energy coupling—that is, using the electrical output of solar and wind systems directly. That's because any time you convert one form of energy to another, energy is lost. Take the hydrogen storage idea as an example. In this system, electrical energy is required to generate hydrogen and oxygen gas from water. It also takes energy to compress these gases to store them. Hydrogen and oxygen then have to be combined in the fuel cell to create new electricity. This process, of course, is not 100% efficient, either.

Creating and storing energy in intermediate forms, like hydrogen, loses energy. It makes much more sense to use the electricity required to split water directly than to convert it into some other form and then convert it back to electricity. Each step in conversions such as these loses a lot the initial energy you started out with. Figure 1.10 shows an example, comparing the efficiency of generating electricity versus using electricity directly to power an electric car. As you can see, it is three to four times more efficient to use electricity directly than to convert it to a storage form, then convert it back. It's the laws of physics (notably the Second Law of Thermodynamics) at play.

As in residential systems, electricity for mass consumption can be supplied on cloudy days or at night by other renewable energy resources, such as wind, hydropower, or biomass (Figure 1.11). For example, commercial wind farms in Iowa and

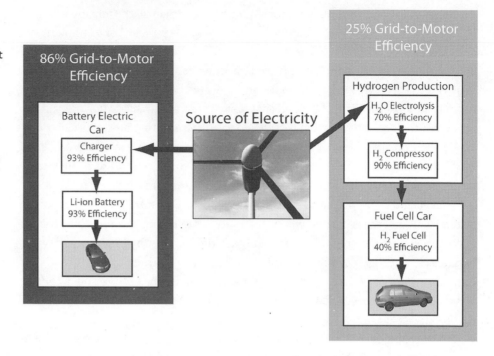

FIGURE 1.10. Efficiency of Direct Solar Vs. Hydrogen Stored Solar. The most fundamental problem with storing electricity in the form of hydrogen, then converting it back to electricity in hydrogen fuel cells is that it involves multiple energy conversions. Each conversion reduces the overall output of the system, as shown here in a system designed to produce electricity to power electric cars. Credit: Anil Rao.

downstate Illinois could provide electricity to Chicago to supplement solar systems in the summer and winter. That's because the winds often blow when the Sun is behind clouds (during storms, for instance). Hydroelectric plants and biomass facilities could also be used to ensure a continuous supply of renewable energy in an electric system finely tuned to switch from one energy source to another. In Canada, for example, hydroelectric facilities are treated as a *dispatchable energy resource*, much like natural gas is today in the United States. That is, they can be turned on and off as needed to meet demand. Such systems could be used to supplement electricity when demand exceeds the capacity of commercial solar systems or at night.

Shortfalls can be offset on a local or a regional scale by transferring electricity from areas of surplus solar and/or wind production to areas of insufficient electrical production. In Colorado, for instance, on cloudy days, wind farms in the northeastern and southeastern parts of the state could supplement locally produced solar electricity along the Front Range, providing electricity to Denver, Colorado Springs, and Fort Collins. The often-sunnier western slope (Grand Junction area) might also provide additional electricity to the Front Range of Colorado from surpluses. If more electricity is needed, it could be shipped via the electrical grid from neighboring Wyoming, Nebraska, and Kansas (Figure 1.12).

FIGURE 1.11. The author's boys, Skyler and Forrest, check out a large wind turbine at a wind farm in Canastota, New York. Wind farms like this one are popping up across the nation—indeed, across the world—producing clean, renewable electricity to power our future. Credit: Dan Chiras.

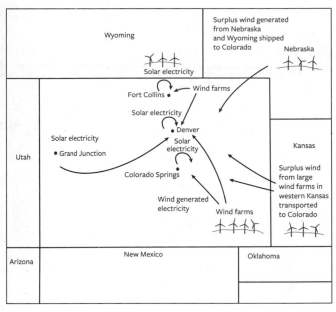

FIGURE 1.12. Electricity can be transported from one state or province to another via the electrical grid, a network of high-voltage transmission lines. As illustrated here, surplus electricity from solar and wind energy systems can be transferred to neighboring states, helping to create a steady supply of electricity through rain and shine. Credit: Forrest Chiras.

Integrating Renewables into America's Electrical Grid

European nations such as Denmark, Germany, and Spain are successfully integrating renewable energy into their electrical grids—and on a large scale. While the electrical grid in the United States is not as finely tuned, efforts are underway to improve and expand the grid, making it more amenable to renewable energy, like solar. Improved weather forecasting and better integration of data on the projected output of large commercial solar and wind systems with projections of customer demand could assist utilities in meeting customers' demands using a mixture of conventional and renewable fuels. Weather projections, for example, allows utilities to step up solar or wind production and scale back production from coal or other fossil fuels or vice versa. Today, thanks to these and other changes, many states are successfully integrating small residential and large commercial solar electric and wind systems with conventional sources of electricity. Further refinements will result in an even greater contribution from renewables—and a cleaner, more sustainable future.

FIGURE 1.13. Solar arrays can be mounted close to the roof to reduce their visibility. Like the arrangement shown here, arrays can be supported by aluminum rails, usually about 6 inches (15 cm) above the surface of the roof (the gap helps cool them in the summer). While attractive, arrays installed this way tend to produce less electricity than pole-mounted or ground-mounted PV arrays, which stay much cooler on hot summer days. Credit: Rochester Solar Technologies.

Solar's availability and variability can also be offset by biomass or natural gas-fired power plants and newer coal-fired plants that burn pulverized coal. Both can be started or stopped, or throttled up or down, to provide electricity at nearly a moment's notice.

Even though solar energy is not available 24 hours a day, there are ways to overcome this problem. That said, we should point out that even though solar energy is variable, it is not unreliable. Just like wind energy, you can count on a certain amount of solar energy each year. With smart planning, forecasting, and careful design and integration of new and existing sources of electricity, we can meet a good portion of our electrical needs from this seemingly capricious resource.

Aesthetics

While many of us view a solar electric array as a thing of great value, even beauty, some don't. Your neighbors, for instance, might think that a PV system detracts from the beauty of the neighborhood. Because not everyone views a PV system the same way, some neighborhood associations have banned PV systems.

Ironically, those who object to solar electric systems rarely complain about other visual blights like cell phone towers, electric transmission lines, and billboards. One reason that these common eyesores draw little attention is that they have been present in our communities for decades. We've grown used to ubiquitous electric lines and cell phone towers. But PV arrays are relatively new, and people aren't used to them yet.

Fortunately, there are ways to mount a solar array so that it blends seamlessly with the roof. As you'll learn in Chapter 8, solar modules can be mounted six inches (15 cm) or so off, and parallel to, the roof surfaces (Figure 1.13). Manufacturers are also producing modules with black frames. They are much less conspicuous than the traditional silver-framed modules. Installations such as these help make solar more appealing to more people. There's also a solar product called PV laminate that is applied directly to certain types of metal roofs, known as *standing seam metal roofs*, resulting in an even lower-profile array (Figure 1.14). Solar arrays can sometimes be mounted on poles or racks anchored to the ground that can be placed in sunny backyards—out of a neighbor's line of sight (Figure 1.15).

Cost

For years, the biggest disadvantage of solar electric systems has been their cost. Even in the last edition of this book, which was published in 2009, costs were still rather high, at about $4.00 per watt just for the modules. Those of us who advocated for solar electricity had to appeal to people's sense of right and wrong and to the external economic costs—that is, the environmental costs—of conventional power. I'm delighted to report that those days are over. As noted earlier, solar electricity is now often cost competitive—and sometimes cheaper—over the long haul than electricity from conventional sources, even in rural areas. I'll show you the proof of this assertion in Chapter 4.

FIGURE 1.14. This product, known as PV laminate, is a plastic-coated flexible material that adheres to standing seam metal roofs. It's best applied to new roof panels *before* they are installed on the building. PV laminate is not as sensitive to high temperatures as conventional silicon-crystal modules used in most solar systems. The company that first developed this product, Uni-Solar is no longer in business, but other companies produce them. Credit: Uni-Solar.

FIGURE 1.15. Ground-mounted solar arrays like this one accommodate numerous modules and can be oriented and angled to maximize production. Because they allow air to circulate freely around the modules, systems such as these stay cooler than many roof-mounted arrays and thus tend to have a higher output. Credit: Dan Chiras.

Solar electricity makes sense in most areas with a decent amount of sunshine. In those areas with lots of sunshine and high electricity costs, like southern California, it makes even more sense. Solar electricity may also make sense for those who are building a home more than a few tenths of a mile from a power line. That's because utility companies often charge customers a gargantuan fee to connect their power lines. You could, for instance, pay $20,000 to connect to an electric line, even if you are building a home 0.2 miles from a power line. Bear in mind, too, that these line extension fees don't pay for a single kilowatt-hour of electricity. They only cover the cost of the transformer, poles, wires, electrical meter, and installation. A homeowner pays the cost of this connection up front or, more commonly, prorated over many years, with the extra amount showing up on the monthly electrical bill.

Another common complaint about solar electricity is that it requires costly, inefficient batteries. Nothing could be further from the truth. Most PV systems installed these days require no batteries. These systems produce electricity that feeds active loads in our homes (that is, things like our TVs and lights). When the system produces more electricity than is needed, the surplus is backfed onto the grid, running the customer's meter backward. However, when the homeowner needs that electricity back, it is his or hers for free—within the billing cycle. (More on this in Chapter 5.) So, batteries are generally not needed.

The Advantages of Solar Electric Systems

Although solar electricity, like any fuel, has some downsides, they're clearly not insurmountable and are outweighed by many advantages. One of the most important advantages is that solar energy is an abundant, renewable resource—one that will be with us for hundreds of thousands of years. While natural gas, oil, coal, and nuclear fuels are limited and on the decline, solar energy will be available to us for about 1 billion years.

Solar energy is a clean resource, too. By reducing the world's reliance on coal-fired power plants, solar electricity could help us reduce our contribution to a host of environmental problems, among them acid rain, global climate disruption, habitat destruction, species extinction, and cropland loss caused by desertification. Solar electricity could even replace costly, risky nuclear power plants. Solar energy could help us decrease our reliance on declining and costly supplies of fossil fuels like coal, natural gas, and oil. (Although very little electricity in the United States comes from oil, electricity generated by solar electric systems could someday be used to power electric or plug-in gas-electric hybrid vehicles [Figure 1.16] like the Ford Fusion Energi.) And, although the production of solar electric systems does have its impacts, all in all it is a relatively benign technology compared to fossil-fuel and nuclear power plants.

Another benefit of solar electricity is that, unlike oil, coal, natural gas, and nuclear energy, the fuel is free. Moreover, solar energy is not owned or controlled by hostile foreign states or one of the dozen or so major energy companies that dictate global energy policy. Because the fuel is free and will remain free, solar energy can provide a hedge against inflation, fueled in part by ever-increasing fuel costs. That's why solar is popular among many businesses and some school districts. Solar is a one-time investment for a lifetime of inexpensive electricity not subjected to annual price hikes.

An increasing reliance on solar and wind energy could also ease political tensions worldwide. Solar and other renewable energy resources could alleviate the perceived need for costly military operations aimed at stabilizing (controlling the politics of) the Mideast, a region where the largest oil reserves reside. Because the Sun is not owned or controlled by any nation in the Middle East, we'll never fight a war over solar or other renewable energy resources. Not a drop of human blood will be shed to ensure the steady supply of solar energy to fuel our economy—or at least, I hope not.

FIGURE 1.16. Plug-in hybrids, like this one, and electric cars will very likely play a huge role in personal transportation in the future. Electric cars with longer-range battery banks could be used for commuting and for short trips under 200 miles, while plug-in hybrids could be used for commuting and long-distance trips. Credit: Calcars.org.

Yet another advantage of solar-generated electricity is that it uses existing infrastructure (the electrical grid). A transition to solar electricity could occur fairly seamlessly.

Solar electricity is also modular. (That's one reason why solar "panels" are called "modules.") That is, you can build a system over time. If you can only afford a small system, you can start small and expand your system as money becomes available. Expandability has been made easier by the invention of *microinverters*, small inverters that are wired to each solar module in a PV system. (For years, most solar systems were designed with one inverter, known as a *string inverter*. If you wanted to expand your system, you'd buy more modules and have to buy a larger, more expensive inverter. With microinverters, a homeowner can add one module and one microinverter at a time. It is much more economical to expand this way.)

Solar electricity could provide substantial economic benefits for local, state, and regional economies. Moreover solar electricity does not require extensive use of water; reliance on water is an increasing problem for coal, nuclear, and gas-fired power plants, particularly in the western United States and other arid regions.

On a personal level, solar electric systems offer considerable economic savings over their lifetime, a topic discussed in Chapter 4. They also create a sense of pride and accomplishment, and they generate tremendous personal satisfaction. A 2015 report by the Lawrence Berkeley National Laboratory on home sales in eight states from 2002 to 2013 showed that solar electric system boosted selling prices. Sales prices on homes with PV systems were about $4.17 per watt higher than comparable "solarless" homes. If you had a 10-kW PV array on your home, you would have made an additional $41,700. That's amazing, especially when you take into account that the cost of a PV system nationwide is currently $3.46 per watt, installed. That system would cost $34,600 up front. Had you availed yourself to the 30% federal tax credit, the cost would have been $24,200. So not only would you have you doubled your money, the systems would have generated a ton of electricity free of charge.

Purpose of this Book

This book focuses primarily on solar electric systems for homes and small businesses. It is written for individuals who aren't well versed in electricity. I'll teach you much of what you need to know about electricity. Rest assured, you don't need a master's degree in electrical engineering to understand this material.

In this book, I've strived to explain facts and concepts clearly and accurately, introducing key terms and concepts as needed, and repeating them as often as prudent to make my points clearly and accurately. My overarching goal is always to create a very user-friendly book.

When you finish reading and studying the material in this book, you'll know an amazing amount about solar energy and solar electric systems. You will have the knowledge required to assess your electrical consumption and the solar resource at your site. You will also be able to determine if a solar electric system will meet your needs and if it makes sense for you. You will know what kind of system you should install, and you'll have a good working knowledge of the key components of PV systems. In keeping with my long-standing goal of creating savvy and knowledgeable buyers, this book will help you know what to look for when shopping for a PV system or an installer. You'll also know how PV systems are installed and what their maintenance requirements are.

I should point out, however, that this book is not an installation manual. When you're done reading, you won't be qualified to install a solar electric system. Even so, this book is a good start. You'll understand much of what's needed to design and install a system or launch a career in PVs. Many people who have read this book and taken my workshops have gone on to careers in solar electricity.

If you choose to hire a solar energy professional to install a system, you'll be thankful you've read and studied the material in this book. The more you know, the more informed input you will have into your system design, components, siting, and installation—and the more likely you'll be happy with your purchase.

This book should also help you develop realistic expectations. Knowing the shortcomings and pitfalls of solar electric systems helps us avoid mistakes and prevent disappointments that are often fueled by unrealistic expectations. By the same token, the more you know about these systems, the more likely you are to install an efficient high-output system. Paying attention to small details can result in huge increases in output.

Organization of this Book

Let's begin our exploration of solar electricity. We'll start in the next chapter by studying the Sun and solar energy. I will discuss important terms and concepts such as average peak sun hours. You'll learn why solar energy varies during the year, and how to calculate the proper orientation and tilt angle of your array to achieve optimal performance.

In Chapter 3, we'll explore solar electricity—the history of PVs, the types of solar cells on the market today, their efficiency, how solar cells generate electricity, and what new solar electric technologies are being developed.

In Chapter 4, we will explore the feasibility of tapping into solar energy to produce electricity at your site. I'll tell you how to assess your electrical energy needs and determine the size of the solar system you'll need to meet them. You'll also learn why it is so important to make your home as energy efficient as possible before you install a solar electric system.

In Chapter 5, we'll examine three types of residential solar electric systems: (1) off-grid, (2) batteryless grid-tied, and (3) grid-connected with battery backup. You'll learn the basic components and the pros and cons of each system. We will also examine hybrid systems—for example, wind-PV systems. You will see how they complement each other and can provide a reliable, year-round supply of electricity. That said, you will also learn whether it makes sense to hybridize or simply expand your solar system. In this chapter, you will also learn the ins and outs of connecting to the electric grid. And I will discuss the sometimes-confusing process called *net metering*. I'll also show you ways to make solar electricity affordable. And, we'll explore ways to expand a solar electric system and ways to economically add batteries to a grid-tied system. This is a strategy that many of us are contemplating as utility companies are starting to charge PV system owners more for electricity they buy from the grid. (More on that revolting development shortly!)

Chapter 6 introduces readers to inverters, string inverters, and microinverters, vital components of most solar electric systems. You'll learn what they do and how they operate. I'll also discuss some recent innovations in inverter design and provide some shopping tips—ideas on what you should look for when buying an inverter.

In Chapter 7, I'll tackle storage batteries for off-grid solar or grid-connected systems with battery backup. You will learn about the types of batteries you can install, how to install batteries correctly, battery maintenance, and ways to reduce battery maintenance. I will point out common mistakes people make with their batteries—and there are many—and give you ways to avoid making those same, often costly mistakes. You will also learn about battery safety and how to size a battery bank for a PV system. I'll finish with a discussion of charge controllers, an essential component of virtually all battery-based PV systems. We'll examine some recent innovations in charge controllers and explore backup generators, providing information on what your options are and what to look for when buying a generator.

In Chapter 8, I'll provide an overview of the steps in a solar electric system installation—so you know what to expect. I'll also discuss system maintenance. You'll learn various mounting options—for example, pole-mounts and roof racks. We'll explore pole-mounted racks that enable PV arrays to track the Sun from sunrise to sunset and discuss the economics of this option.

In Chapter 9, we'll explore a range of issues such as permits, covenants, and utility interconnection. I'll discuss whether you should install a system yourself or hire a professional and, if you choose the latter, how to locate a competent installer.

Finally, as all of my books do, this book includes a comprehensive resource guide. It contains a list of books, articles, videos, associations, organizations, workshops, and websites of manufacturers of the components of PV systems.

Understanding the Sun
and Solar Energy

The Sun lies in the center of our solar system, approximately 93 million miles from Earth. Composed primarily of hydrogen and small amounts of helium, the Sun is a massive nuclear reactor. However, it's not the same type of reactor found in the nuclear power plants that some utility companies use to generate electricity. Those are *fission* reactors designed to split atoms of uranium-235 in a controlled fashion. The heat generated in this process boils water to generate steam. Steam spins a turbine that drives a generator that produces electricity.

The Sun is a giant *fusion* reactor. Nuclear fusion occurs in the Sun's core, where intense pressure and heat force hydrogen atoms to fuse. This results in the formation of slightly larger helium atoms and immense amounts of energy. This energy migrates to the surface of the Sun, taking up to a million years to make it there. When this energy reaches the surface of the Sun, it radiates into space, primarily as light and heat.

Solar radiation streaming into space strikes the Earth, warming and lighting our planet and fueling virtually all life. What's remarkable, though, is that the Earth receives only a tiny fraction of the Sun's output—less than one billionth of the energy that radiates from its surface. Although our allotment is small, "the solar energy received each year by the Earth is roughly…10,000 times the total energy consumed by humanity," according to French energy expert, Jean-Marc Jancovici. To replace *all* the oil, coal, gas, and uranium currently used to power human society with solar energy, we'd only need to capture 0.01% of the energy of the sunlight striking the Earth each day. Forty minutes worth of sunlight is equivalent to all the energy consumed by human society in a year!

According to the US Department of Energy's National Renewable Energy Laboratory (NREL), to generate the electricity the United States consumes, we'd only need to install PVs on 7% of the land surface area currently occupied by cities and homes. We could achieve this goal by installing PVs on rooftops of homes, factories, and office buildings; over parking lots; and on the south sides of buildings where solar arrays could serve as awnings (Figure 2.1 a–c). We wouldn't have to appropriate a single acre of new land to make PV our primary energy source!

In this chapter, we'll explore the Sun and solar radiation. We'll examine key concepts including irradiance, irradiation, and peak sun hours. We'll study the difference between direct and diffuse radiation and explore key concepts such as azimuth angle, altitude angle, and tilt angle. An understanding of these concepts is important for mounting PV arrays to maximize energy output from a PV array.

Understanding Solar Radiation

The Sun releases immense amounts of energy from its surface every day. This output, known as *solar radiation*, consists primarily of *electromagnetic*

FIGURE 2.1. Solar electric systems can be mounted in ways that don't require big structural changes. (*a*) This copy shop in Union, Missouri, for example, installed modules on a rack south of the building, providing shade in the summer and a great place to park one's car, albeit very carefully, on a hot summer afternoon. (*b*) The roofs of commercial buildings, like this Pro-lube Express in St. Joseph, Missouri, provide plenty of square footage that can be used to mount large solar arrays. Imagine the size of the solar arrays that could be mounted on large factories and box stores. (*c*) This array in St. Louis was made to perform double-duty, serving as both a source of electricity and as an awning to provide shade in the summer. Credit: Dan Chiras.

radiation. As shown in Figure 2.2, the Sun's electromagnetic radiation ranges from high-energy, short-wavelength gamma rays to low-energy, long-wave radiation radio waves. In between these extremes, starting from the short-wave end of the spectrum are X-rays, ultraviolet radiation, visible light, and heat (infrared radiation).

While the Sun releases numerous forms of electromagnetic radiation, most of it (about 40%) consists of infrared radiation (commonly called heat) and visible light (about 55%). Traveling at a speed of 186,000 miles per second, solar energy takes a little over 8 minutes—8.3 minutes, to be more precise—to make its 93-million-mile journey to the Earth. Solar electric (PV) modules capture only a portion of this energy—notably, the energy contained in the visible and lower end of the infrared portions of the spectrum. We call the latter *near infrared radiation.*

Electromagnetic radiation from the Sun travels virtually unimpeded through space until it encounters the Earth's atmosphere. In the outer portion of the atmosphere (known as the *stratosphere*), solar radiation encounters a region known as the *ozone layer.* As most readers probably know, ozone molecules (O_3) absorb much (99%) of the incoming ultraviolet radiation, dramatically reducing our exposure to

Solar Variation

The Sun's intensity varies by region. The Desert Southwest, for instance, is blessed with sunlight. The Midwest receives less but still plenty of sunlight to make a solar electric system worthwhile. On an annual basis, for example, Kansas City receives only about 25% less sunlight than Phoenix. Buffalo, New York, one of the cloudiest regions in the United States, receives about 50% less sunlight than Kansas City. Solar electricity works in all regions, although the size of systems must be increased in cloudier regions to make up for reduced sunshine.

Electromagnetic Spectrum

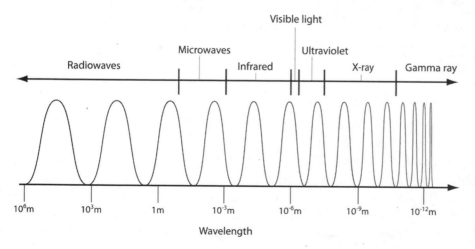

FIGURE 2.2. Electromagnetic Spectrum. The Sun produces a wide range of electromagnetic radiation, from gamma rays to radio waves, as shown here. PV cells convert visible light into electric energy, specifically direct current electricity. Credit: Anil Rao.

this potentially harmful form of solar radiation. As the rest of the sunlight passes through the lower portion of the atmosphere (the troposphere), it encounters clouds, water vapor, pollution, and dust. These may absorb the Sun's rays or reflect them back into space, reducing the amount of sunlight striking the Earth's surface.

PVs and Solar Absorption

The vast majority of solar cells in use today are made from crystal silicon. These cells produce DC electricity from solar energy in the visible and lower end of the infrared portion of the electromagnetic spectrum—radiation in the 300 to 1,100 nanometer range. (A nanometer [nm] is one billionth of a meter.) Some solar modules are made from thin layers of photo-reactive materials sprayed on metal, glass, or plastic. These are known as *thin-film modules*. Thin-film solar modules generally respond to a much narrower range—from 300 to slightly under 600 nm—although newer designs consisting of multiple layers of thin-film material can increase the upper end of the range to over 1,100 nm.

Measuring Irradiance

Scientists use a device known as a *pyranometer* to measure solar irradiance. Pyranometers measure both direct and diffuse radiation. Direct radiation consists of sunlight that travels directly through the atmosphere. Diffuse radiation is sunlight energy that bounces off clouds. It's the type of light you experience on a cloudy day. A shading device placed over a pyranometer allows it to measure only diffuse radiation. Inexpensive pyranometers can be mounted alongside solar arrays to monitor solar input and gauge the performance of a system.

Irradiance

The intensity of solar energy striking a surface is known as *irradiance*. Irradiance is measured in watts per square meter (W/m^2). Solar irradiance measured just before the Sun's radiation enters the Earth's atmosphere is about 1,366 W/m^2. On a clear day, nearly 30% of the Sun's radiant energy is absorbed or reflected by dust and water vapor in the Earth's atmosphere. (Absorbed sunlight is converted to heat.) By the time the incoming solar radiation reaches a rooftop solar array, the 1,366 W/m^2 measured in outer space has been winnowed down to 1,000 W/m^2. (As you shall soon see, this is the highest irradiation encountered at most locations throughout the world and is called *peak sun*.)

Even though the Sun experiences long-term (11-year) cycles, during which its output varies, over the short term, irradiance remains fairly constant. On Earth, however, solar irradiance (the intensity of sunlight) varies during daylight hours at any given site. At night, as you'd expect, solar irradiance is zero. As the Sun rises, irradiance increases slowly but surely, plateauing between 10 AM and 2 PM. Afterward, irradiance slowly decreases, falling once again to zero at night.

Changes in irradiance are determined by the angle of the Sun's rays—that is, the angle at which the Sun's rays strike the surface of the Earth. This changes by the minute as the Earth rotates on its axis. The angle at which the Sun's rays strike the Earth affects two important factors: energy density (Figure 2.3) and the amount of atmosphere through which sunlight must travel to reach the Earth's surface (Figure 2.4).

As shown in Figure 2.3, early morning and late-day sunlight, that is, low-angled sunlight, delivers much less energy per square meter than high-angled sunlight, resulting in decreased irradiance at these times. As the Sun makes its way across the sky and the sunlight streams in from directly above irradiance increases.

The angle at which the Sun's rays pass through the atmosphere also affects the amount of atmosphere through which sunlight passes, as shown in Figure 2.4. The more atmosphere, the more filtering of incoming solar radiation, the less sunlight makes it to Earth, and the lower the irradiance. I'll elaborate on this concept shortly.

Irradiation

Irradiance is a measure of instantaneous power. Although it is important, what most solar installers need to know is irradiance over time—the amount of energy they can expect to capture each day. Irradiance over a period of time is referred to as *solar irradiation* or simply *irradiation*. Irradiation is expressed as watts per square meter striking the Earth's surface (or a PV module) for some specified period of time—for example, per hour. As you may recall from Chapter 1, power is measured in watts. One hundred watts of sunlight striking a square meter of surface for one hour is expressed as 100 watt-hours per square meter. Solar radiation of 500 watts of solar energy striking a square meter for an hour is 500 watt-hours per square meter. As you shall soon see, solar installers are more interested in total daily irradiation.

To help keep these terms straight, remember that irradiance is an instantaneous measure of

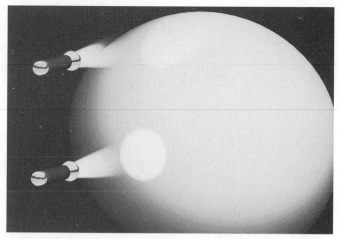

FIGURE 2.3. Lower Energy Density from Low-Angled Winter Sun. Surfaces perpendicular to the incoming rays absorb more solar energy than surfaces not perpendicular, as illustrated here. Credit: Anil Rao.

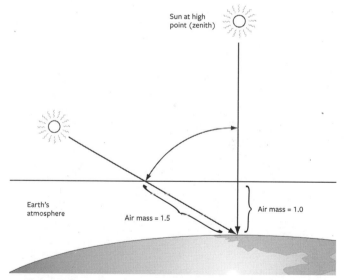

FIGURE 2.4. Atmospheric Air Mass and Radiation Reaching the Earth's Surface. Early and late in the day, sunlight travels through more air in the Earth's atmosphere, which decreases the amount of energy reaching a solar electric array, decreasing its output. Maximum output occurs when the Sun's rays pass through the least amount of atmosphere, at solar noon. Three-quarters of the daily output from a solar array occurs between 9 AM to 3 PM. Credit: Anil Rao.

power. Irradiation, on the other hand, is a measure of power over time. Physicists refer to this as energy. I help students keep the terms straight by likening irradiance to the speed of a car. Like irradiance, speed is an instantaneous measurement. It tells us how fast a car is moving at any moment. Irradiation is akin to the distance a vehicle travels. Distance, of course, is determined by multiplying the speed of a vehicle by time it travels at a given speed. It is a quantity, not a rate. The same holds true for irradiation.

Figure 2.5 illustrates the concepts graphically. As shown, irradiance is the single black line in the graph—the number of watts per square meter at any given time. The area under the curve is the total solar irradiance during a given period, in this case, a single day. The total irradiance in a day is irradiation.

Solar irradiation is useful to professional installers and do-it-yourselfers when sizing PV systems—that is, determining the size of the system needed to meet the electrical demands of customers. Solar irradiation is also a useful measure for the few utilities that base solar rebates on the projected electrical production of their customers' PV systems.

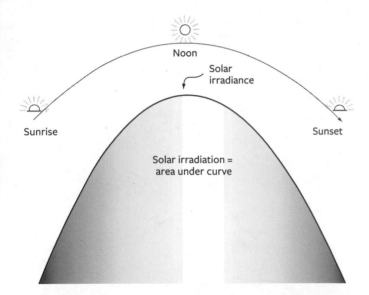

FIGURE 2.5. Graph of Solar Irradiance and Solar Radiation. This graph shows both solar irradiance (watts per square meter striking the Earth's surface) and solar irradiation (watts per square meter per day). Note that irradiation is total solar irradiance over some period of time, usually a day. Solar irradiation is the area under the curve. Credit: Anil Rao.

Direct vs. Diffuse Radiation

Now that you understand irradiance and irradiation, let's take a closer look at solar radiation to see what happens to it as it passes through the atmosphere. As shown in Figure 2.6, some sunlight entering the Earth's atmosphere passes through the sky unimpeded. It reaches the Earth's surface directly—that is, without being blocked by clouds, dust, pollutants, or water vapor. This is referred to as *direct radiation*. The parallel rays of sunlight in direct radiation produce distinct shadows. The remainder of the Sun's radiation is either absorbed or scattered (reflected) by clouds, water molecules, pollutants, and dust suspended in the atmosphere. Visible light in solar radiation absorbed by the Earth's atmosphere is converted to heat. Some heat radiates to Earth; some radiates into outer space. Visible light reflected off clouds, dust, and so on, is dispersed in many directions, as shown in Figure 2.6. This is diffuse radiation. As illustrated, some diffuse radiation is reflected into outer space; the rest radiates down to Earth. Unlike direct radiation, whose rays run parallel to one another, diffuse radiation arrives at many angles. As a result, it produces no shadows.

On a clear day, direct radiation accounts for 80% to 90% of the sunlight striking the Earth's surface. Diffuse radiation accounts for the rest. On a cloudy day, 100% of the sunlight striking the Earth's surface is diffuse radiation.

The relative proportion of direct and diffuse radiation striking the Earth's surface is influenced by a number of factors. One of the most important is the amount of atmosphere through which solar radiation must pass to reach us. The amount of atmosphere, in turn, affects the amount of absorption and scattering that occur. On any given day, for instance, early morning and late-day sunrays travel through more atmosphere than when the Sun is higher in the sky (Figure 2.4). The more atmosphere through which sunlight must travel, the more absorption and scattering take place. The more absorption and scattering that occurs, the less direct radiation that strikes the Earth's surface. The less direct radiation, the lower the irradiance and the lower a PV array's electrical output.

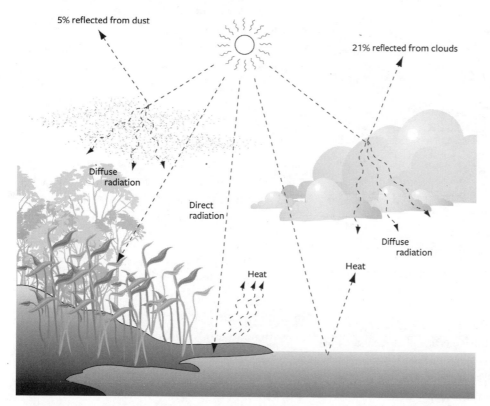

FIGURE 2.6. Diffuse and Direct Radiation. Sunlight streaming through the Earth's atmosphere may pass through unimpeded. This is known as direct radiation. Or, it may be scattered by dust or clouds. Scattered light is known as diffuse radiation. Credit: Anil Rao.

Time of day is not the only factor that affects irradiance and irradiation. Elevation also plays an important role. All things being equal, high-elevation regions receive more direct sunlight than low-elevation regions because there's less atmosphere to absorb and scatter incoming solar radiation.

The vast majority of solar modules absorb both direct and diffuse radiation and convert it into electricity. That's why a solar array will continue to produce electricity on a cloudy day—although at a lower rate than on a sunny day.

To increase the efficiency of solar electric systems, some companies install *concentrating solar collectors*. Concentrating solar collectors are equipped with concentrating lenses or mirrors that focus solar energy onto the PV cells, greatly increasing their output (Figure 2.7). Although they are considerably more efficient than standard flat modules, their use is restricted to large commercial arrays—arrays that produce electricity commercially. Some roof racks for flat roofs on commercial buildings come equipped with polished "mirrors" that reflect additional solar energy onto the modules, which enhances production.

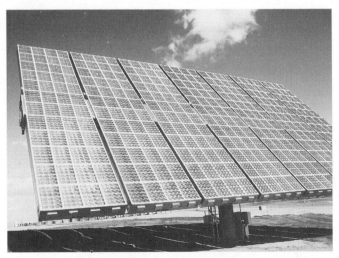

FIGURE 2.7. Concentrating Solar Cells. Some concentrating solar systems like the one shown here use inexpensive plastic lenses to focus (concentrate) sunlight onto high-efficiency PV cells, minimizing the size of the solar cell required and increasing output. This photo is of the Amonix High Concentration Photovoltaic (HCPV) Solar Power Generator, which has a generating capacity of 53kW and stands about 50 feet tall and 77 feet wide. This array is designed for utility-scale solar power generation. Credit: Amonix.

Peak Sun, Peak Sun Hours, and Insolation

Another measurement installers use when designing PV systems is *peak sun*. As just noted, peak sun is the maximum solar irradiance available at most locations on the Earth's surface on a clear day—1,000 W/m² (or 1 kWh/m²). One hour of peak sun is known as a *peak sun hour*. Four hours of peak sun are 4 peak sun hours. Figure 2.8 shows a map that shows the daily average peak sun hours in the United States and Canada measured as kW/m²/day. (For best viewing, find a map like this, but in full color on the internet.)

Peak sun hours is a measure of solar irradiation (watts per m² per day) and is also referred to as solar *insolation* or simply insolation—although these terms are not commonly used in the solar business today. Average peak sun hours per day are used to compare solar resources of different regions. The annual average in southern New Mexico and Arizona, for instance, is 6.5 to 7 peak sun hours per day. In Missouri, it's 4 to 4.5 peak sun hours per day. In cloudy British Columbia, it's about 4. Remember, though, these are

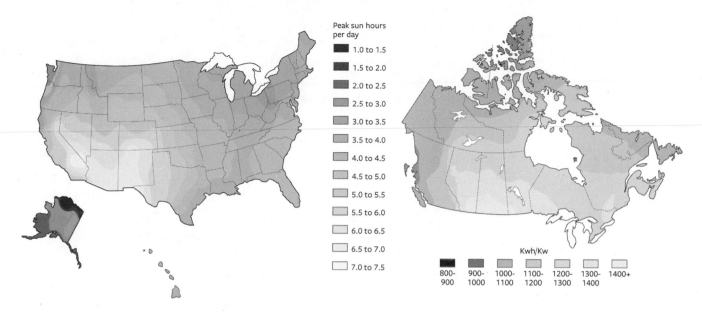

Peak sun hours per day

- 1.0 to 1.5
- 1.5 to 2.0
- 2.0 to 2.5
- 2.5 to 3.0
- 3.0 to 3.5
- 3.5 to 4.0
- 4.0 to 4.5
- 4.5 to 5.0
- 5.0 to 5.5
- 5.5 to 6.0
- 6.0 to 6.5
- 6.5 to 7.0
- 7.0 to 7.5

Kwh/Kw

| 800-900 | 900-1000 | 1000-1100 | 1100-1200 | 1200-1300 | 1300-1400 | 1400+ |

averages. In the summer in virtually all locations, peak sun hours are considerably greater than in the winter.

Installers determine peak sun hours by consulting detailed state maps, tables, or various web sites. They use average peak sun hours to size solar arrays—that is, to determine how large of an array will be required to meet a customer's needs. They also use it to estimate the annual output of existing arrays to troubleshoot—for example, if a solar array appears not to be performing as well as expected. It's important to note that the average peak sun hours per day for a given location doesn't mean that the Sun shines at peak intensity during that entire period. In fact, peak sun conditions—solar irradiance equal to 1,000 watts/m²—will very likely only occur two to four hours a day. So how do scientists calculate peak sun hours per day?

Peak sun hours are calculated at a given location by determining the total irradiation (watt-hours per square meter) during daylight hours and dividing that number by 1,000 watts per square meter. On a summer day, for example, let's assume that solar irradiance averages 600 watts per square meter for 4 hours, 800 watts per square meter for 4 hours, and 1,000-watts per square meter for 2 hours. Four hours at 600 W/m² is 2,400 watt-hours/m². Four hours at 800 W/m² is 3,200 watt-hours/m². Two hours at 1,000 W/m² is 2,000 watt-hour/m². Add them up, and you have 7,600 watt-hours/m². That's the daily solar irradiation. The number of peak sun hours is 7,600 watt-hours/m² divided by 1,000. In this case, we'd have 7.6 peak sun hours. Even though average peak sun hours is 7.6, solar irradiance only

FIGURE 2.8. Map of the United States and Canada showing Average Peak Sun Hours per Day. These maps shows average peak sun hours per day for the U.S. and Canada. This data is used to estimate the annual electrical output of a solar array. Credit: Anil Rao.

reached 1,000 W/m² (peak sun) for 2 hours. To help you think about this, bear in mind that two hours of sunlight at 500 W/m² is equal to one peak sun hour. Four hours at 250 W/m² is also one peak sun hour.

The Sun and the Earth: Understanding the Relationships

To understand solar electric systems, you also need to understand the relationships between the Earth and Sun. In this section, you'll study the Earth's tilt and the Earth's orbit around the Sun and how they affect the amount of sunlight striking the Earth at different times of the year. You'll also learn about the altitude angle and azimuth angle of the Sun, two additional factors that affect the amount of sunlight striking a solar array. This material will help you understand the proper orientation and the angle at which the PV solar modules are mounted to maximize output.

Day Length and Altitude Angle: The Earth's Tilt and Orbit Around the Sun

As all readers know, the Earth orbits around the Sun, completing its sojourn every 365 days. As shown in Figure 2.9, the Earth's orbit is elliptical. As a result, the distance from the Earth to the Sun varies throughout the year. Although many people believe that the distance from the Earth to the Sun determines the seasons, that's not true. In fact, the Earth is closest to the Sun during the winter and farthest from the Sun in the summer. What determines seasons is the amount of sunlight that strikes the Earth. It's determined by the Earth's tilt.

As shown in Figure 2.10, the Earth's axis is tilted 23.5°. The Earth maintains this angle throughout the year as it orbits around the Sun. Look carefully at Figure 2.9 to see that the angle remains fixed—almost as if the Earth were attached to a wire an-

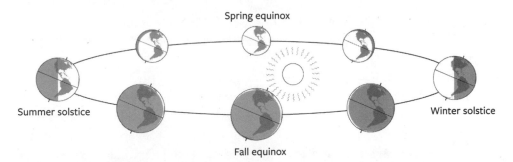

FIGURE 2.9. The Earth's Orbit around the Sun. As any school child can tell you, the Earth orbits around the Sun. Its orbit is not circular, but elliptical, as shown here. Notice that the Earth is tilted on its axis and is farthest from the Sun during the summer. Because the Northern Hemisphere is angled toward the Sun, however, our summers are warm. Credit: Anil Rao.

chored to a fixed point in space. Because the Earth's tilt remains constant, the Northern Hemisphere is tilted away from the Sun during the winter. As a result, the Sun's rays enter and pass through Earth's atmosphere at a very low angle. Sunlight penetrating at a low angle passes through more atmosphere and is therefore absorbed or scattered by more dust, water vapor, and pollutants than sunlight arriving at a steeper angle. This is one reason winters are colder. It is also one reason why solar arrays produce less electricity in the winter.

Surface temperature and irradiance are also lowered because the density of sunlight striking the Earth's surface is reduced in the winter when it strikes at an angle. As shown in Figure 2.3, a surface perpendicular to the Sun's rays absorbs more solar energy than one that's tilted away from it. As a result, low-angled sunlight delivers much less energy per square meter of surface in the winter than it does during summer months.

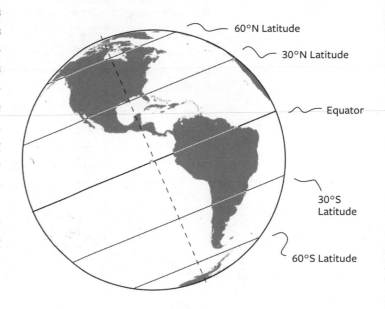

FIGURE 2.10. The Earth's Axis. The Earth is tilted on its axis of rotation, a simple fact with profound implications on solar energy use on Earth. Credit: Anil Rao.

Solar gain is also reduced in the winter because days are shorter—that is, there are fewer hours of daylight during winter months. All three factors combine to reduce the amount of solar energy available to a PV system in the winter. They are also responsible for the cooler temperatures of fall, winter, and early spring, all of which increase the efficiency of solar electric modules. This, in turn, slightly offsets reduced solar gain.

In the summer in the Northern Hemisphere, the Earth is tilted *toward* the Sun, as shown in Figure 2.9 and 2.11. This results in several key changes. One of them is that the Sun is positioned higher in the sky. As a result, sunlight streaming onto the Northern Hemisphere passes through less atmosphere. This reduces absorption and scattering of incoming solar radiation, which increases solar irradiance. Increased irradiance increases the output of a PV array—the energy production in kilowatt-hours. Because a surface perpendicular to the Sun's rays absorbs more solar energy than one that's tilted away from it, the Earth's surface intercepts more energy during the summer as well. Put another way, the high-angled sun delivers much more energy per square meter of surface area than in the winter. Energy density is greater. Moreover, days are also longer in the summer. All these factors increase the output of a PV array, although the warmer days of summer slightly offset these gains (an array's energy output decreases as temperatures increase at any given irradiance level).

Figure 2.12 shows the position of the Sun as it "moves" across the sky during different times of the year as a result of the changing relationship between the Earth and the Sun. This illustration shows that the Sun "carves" a high path across the summer sky. It reaches its highest arc on June 21, the longest day of the year, also known as the *summer solstice.* (*Sol* stands for sun, and *stice* is derived from a Latin word that means to stop.) Figure 2.12 also shows that the lowest arc occurs on December 21, the shortest day of the year. This is the *winter solstice.* The angle between the Sun and the horizon at any time during the day is referred to as the *altitude angle.*

As shown in Figure 2.12, the altitude angle decreases from the summer solstice to the winter solstice. After the winter solstice, however, the altitude angle increases, growing a little each day, until the summer solstice returns. Day length changes

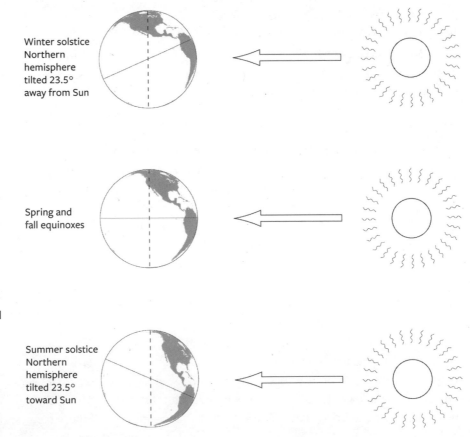

Winter solstice
Northern
hemisphere
tilted 23.5°
away from Sun

Spring and
fall equinoxes

Summer solstice
Northern
hemisphere
tilted 23.5°
toward Sun

FIGURE 2.11. Summer and Winter Solstice. Notice that the Northern Hemisphere is bathed in sunlight during the summer solstice because the Earth is tilted toward the Sun. The Northern Hemisphere is tilted away from the Sun during the winter. Credit: Anil Rao.

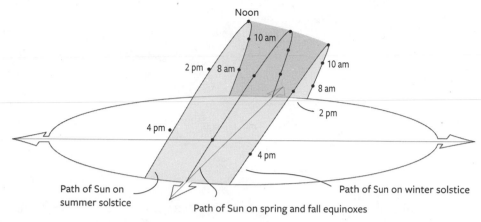

Noon
10 am
2 pm 8 am
10 am
8 am
2 pm
4 pm
4 pm
Path of Sun on
summer solstice
Path of Sun on winter solstice
Path of Sun on spring and fall equinoxes

FIGURE 2.12. Solar Paths. This drawing shows the position of the Sun in the sky during the day on the summer and winter solstices and the spring and fall equinoxes. This plot shows the solar window—the area you want to keep unshaded so the Sun is available for generating solar electricity. Credit: Anil Rao.

along with altitude angle, decreasing about two minutes a day for six months from the summer solstice to the winter solstice, then increasing two minutes a day until the summer solstice arrives once again.

The midpoints in the six-month cycles between the summer and winter solstices are known as *equinoxes*. (The word *equinox* is derived from the Latin words *aequus* [equal] and *nox* [night].) On the equinoxes, the hours of daylight are nearly equal to the hours of darkness. The spring equinox occurs around March 20 and the fall equinox occurs around September 22. These dates mark the beginning of summer and fall, respectively.

The altitude angle of the Sun changes day by day, but it also varies according to the time of day. This change in altitude angle is also determined by the rotation of the Earth on its axis. As seen in Figure 2.12, the altitude angle increases between sunrise and noon, then decreases to zero once again at sunset.

In addition to the change in the altitude angle of the Sun, the Sun's position in the sky relative to the points on the compass changes during the day. For years, solar scientists have referenced the Sun's position in the sky during a day to a fixed point, known as *true south* (defined shortly).

As illustrated in Figure 2.13, true south is assigned a value of 0°. East is +90° and west is –90°. North is 180°. The angle between the Sun and 0° azimuth—or true south (the reference point)—is known as the *solar azimuth angle*. If the Sun is east of true south, the azimuth angle falls in the range of 0° to +180°; if it is west of true south, the azimuth angle falls between 0° and –180°. Like altitude angle, azimuth angle changes as a result of the Earth's rotation on its axis. Take a moment or two to study Figure 2.13 so you are clear on this term.

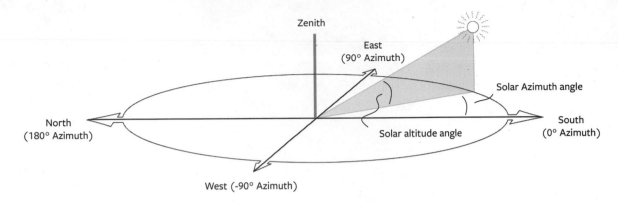

Zenith

East
(90° Azimuth)

Solar Azimuth angle

North
(180° Azimuth)

Solar altitude angle

South
(0° Azimuth)

West (-90° Azimuth)

FIGURE 2.13. Altitude and Azimuth Angle. The altitude angle is the angle of the Sun from the horizon. It changes minute by minute as the Earth rotates. It also changes by day. The azimuth angle is the angle of the Sun from true south. Credit: Anil Rao.

To avoid confusion, increasing numbers of solar installers, authors of books and articles on solar, and software developers have switched from the term azimuth angle to *bearing angle*, a fancy term for points of the compass. In this system, true north is 0°. True south is 180°. East is 90° and west is 270°. If you are at all familiar with compasses, you can readily see why this system is easier to understand.

While azimuth angle changes by the minute, it also shifts by season. That is to say, the azimuth angle or, better yet, bearing angle, of the Sun at any time of day changes from one day to the next day. For example, the bearing angle of the Sun at 10 AM on January 1 differs from the bearing angle on January 30 at 10 AM, as the Sun's path gets longer each day. Check out Figure 2.12 to see how bearing angle changes from one season to the next.

Implications of Sun-Earth Relationships on Solar Installations

Solar modules are typically installed either on poles or racks mounted on the ground or on racks mounted on roofs of buildings. Racks can be fixed (unadjustable) or adjustable.

Ideally, for best year-round production, fixed racks should be oriented to true south (described shortly). The angle at which they are installed, the tilt angle, is fixed. As shown in Figure 2.14, the tilt angle is the angle between the surface of the array and an imaginary horizontal line extending back from the bottom of the array. The ideal tilt angle that most sources recommend is usually equal to the latitude of the site. If you live at 45° north latitude, for instance, most installers suggest tilting the array at a 45° angle. In my experience, however, I've found that for best year-round production the tilt angle of a fixed array should be latitude minus 5°. This recommendation is based on data supplied by NASA's website "Surface Meteorology and Solar Energy" (eosweb.larc.nasa.gov).

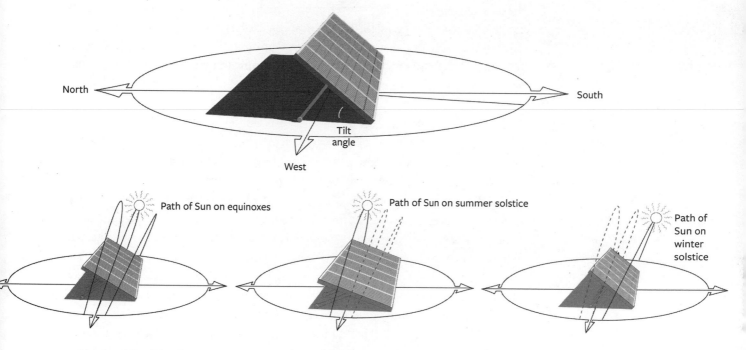

FIGURE 2.14. Tilt Angle. (*top*) The tilt angle of an array, shown here, is adjusted according to the altitude angle of the Sun. The tilt angle can be set at one angle year round, known as the optimum angle, or (*bottom*) adjusted seasonally or even monthly, although most homeowners find this too much trouble. Credit: Anil Rao.

When mounting PV arrays, always be sure to orient them to true south if you live in the Northern Hemisphere and true north if you live in the Southern Hemisphere. True north and south are imaginary lines that run from the North Pole to the South Pole, parallel to the lines of longitude. (True north and south are also known as true geographic north and south.) Magnetic north and south, on the other hand, are lines of force created by the Earth's magnetic field. They are measured by compasses. Unfortunately, magnetic north and south rarely line up with the lines of longitude—that is, they rarely run true north and south. In some areas, magnetic north and south can deviate quite significantly from true north and south. How far magnetic north and south deviate from true north and south is known as the *magnetic declination.*

Figure 2.15 shows the deviation of magnetic north and south—the magnetic declination—from true north and south in North America. You may want to take a moment to study the map. Start by locating your state and then reading the value of the closest isobar. If you live in the eastern United States, you'll notice that the lines are labeled with a minus sign. This indicates a westerly declination—meaning that true south is west of magnetic south. What that means is that when you standing at your site, the compass will point to magnetic south, but true south is west of that line. If you live in western Pennsylvania, for instance, true south will be 10 degrees west of magnetic south.

If you live in the midwestern and western United States, the lines are labeled with a plus sign. This indicates an easterly declination. That is, true north and south lie east of magnetic north and south. When standing anywhere along the 10-degree line, for instance, true south will be 10 degrees east of magnetic south.

To determine the magnetic declination at your site, you can consult the map shown in Figure 2.15. However, for the most accurate indication of magnetic deviation, it is best to contact a local surveyor or a nearby airport. Because magnetic fields can change from year to year, these sources will provide the most accurate information. Be sure to ask whether the magnetic declination is eastly or westerly. Also bear in mind that compass readings at a site may deviate because of local iron ore deposits, metal buildings or even vehicles. So, if you take a magnetic reading too close to a vehicle or a metal barn, the compass may not point precisely to magnetic south. In some cases, it could be way off. Local variations in the magnetic field are called *magnetic deviation*. To avoid magnetic deviation, be sure to move away from metal buildings, metal fences, vehicles, and other large metal objects.

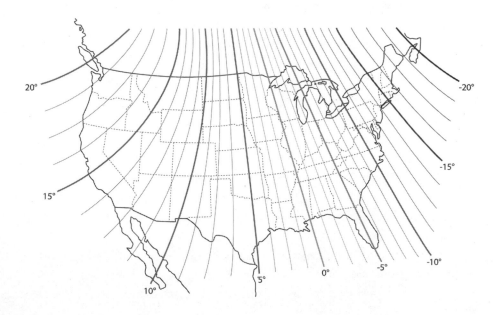

FIGURE 2.15. Map of Magnetic Declination. The isobars on this map indicate magnetic declination in the United States. Note that negative numbers indicate a west declination (meaning true south lies west of magnetic south). Positive numbers indicate an east declination (meaning true south is east of magnetic south). Credit: Anil Rao.

If you think that your site may be compromised by a local deposit of iron ore, a failsafe method of finding true north and south is to head outside on a cloudless night and locate the North Star, Polaris. "The North Star is always either exactly or just a few tenths of a degree from true north," says Grey Chisholm, author of "Finding True South the Easy Way" in *Home Power* (Issue 120). How do you locate it?

The North Star is situated at the end of the handle of the Little Dipper, as shown in Figure 2.16. I've found that the Little Dipper can be a bit difficult to discern. To locate the North Star, follow "the two pointer stars on the outside of the Big Dipper's cup," says Chisholm. A line drawn from them will take you directly to the North Star.

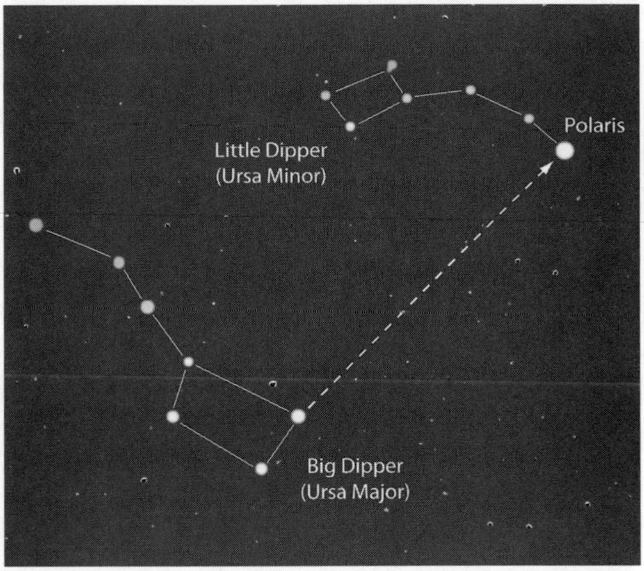

To determine true north and south from the North Star, drive a fence post into the ground, then move south a bit so you are in line with the post and the North Star. Drive a second fence post into the ground at this point. You've now created a line that runs true north and south. Use this line to orient your PV array.

FIGURE 2.16. North Star and Little Dipper. Use the North Star (Polaris) to locate true north. Credit: Anil Rao.

Ideally, a solar array should point directly at the Sun from sunrise to sunset to produce the most electricity. That's because doing so optimizes the energy density of sunlight striking the array, a topic discussed earlier in this chapter. Doing so also ensures sunlight strikes the array head-on—that is, perpendicular to the surface of the array—at all times. This minimizes reflection of sunlight off the surface of the module and ensures maximum absorption. (This is a secondary benefit, and not as significant as maximizing energy density.)

The angle at which sunlight energy strikes a surface of an array—or any object, for that matter—is known as the *angle of incidence*. Technically, the angle of incidence is the angle between incoming solar radiation and a line perpendicular to the surface of the module. When sunlight strikes a module at a 90° angle (perpendicular) the angle of incidence is zero. Increasing the angle of incidence reduces solar irradiance and increases the amount of sunlight that reflects off the array, decreasing output. Ideally, then, the angle of incidence (angle of the incoming solar radiation) should be 0° at all times.

Aligning the array at all times with the Sun to reduce the angle of incidence is possible through the use of automatic trackers. Adjustable racks resemble fixed racks, but are designed so the tilt angle can be changed during the year to increase the output of a PV array. The tilt angle of adjustable racks is manually altered. For best results with the least amount of effort, I recommend changing the tilt angle twice a year, once in the fall and again in the spring. In the fall, tilt angle should be increased by about 15° to more directly align the modules with the low-angled winter sun. In the spring, the tilt angle should be decreased by 15° to capture more energy from the high-angled summer sun.

Some adjustable racks are designed to automatically adjust for the solar azimuth and/or the altitude angle. Racks that automatically adjust to track the Sun from sunrise to sunset (the azimuth angle) are known as *single-axis trackers*. They can increase the output of an array up to 30% in ideal situations.

Tracking arrays that automatically adjust for both the altitude and the azimuth (bearing) angle are known as *dual-axis trackers*. They can increase the output of an array up to 40%, again, in ideal situations.

Although tracking helps improve the output of an array, most PV modules are mounted on fixed racks—set at a specific angle all year round for convenience. Fixed racks can be attached to poles or mounted on racks on the ground or on racks on the roofs of buildings. You will learn more about them in Chapter 8.

Conclusion

Solar energy is an enormous resource that could contribute mightily in years to come. It could heat and cool our homes and power lights and dozens of household electronic devices from dishwashers to television sets. It could even provide power to electric cars used by commuters. To make the most of it, though, a solar system has to be installed properly—that is, oriented properly to take into account the Sun's daily and seasonally changing position in the sky. In the next chapter, we turn our attention to the technologies that capture solar energy: solar cells and solar modules.

CHAPTER 3

Understanding Solar Electricity

Many people view solar electricity as a new or novel technology. Or, if they're familiar with the space program, they may view solar electricity as a by-product of the aerospace industry's efforts to power satellites, work that dates back to the early 1950s. Fact is, the development of solar electricity began well over 100 years ago.

In this chapter, I'll briefly review the history of solar electricity, then describe how PV modules convert sunlight energy to electricity. I'll discuss the types of modules on the market today and some of the newest and most promising PV technologies. In this chapter, I'll also introduce you to some terminology that will be helpful to you if you are a homeowner who wants to install a PV system or an individual looking to start a career in solar electricity.

Brief History of PVs

Like so many ideas whose time has come, solar electricity got its start a long time ago in a series of puzzling and seemingly unimportant discoveries. Edmund Becquerel, a 19-year-old French scientist, began the journey of discovery in 1839 when he immersed two brass plates in a conductive liquid and shined a light on the apparatus. He found that an electrical current was produced.

Because electricity had few uses at the time, his discovery sparked little interest. Becquerel, who went on to contribute mightily to physics, had apparently uncovered a scientific oddity of little value.

Then, in 1873, a British engineer, Willoughby Smith, made another important discovery while working on undersea telegraph cables. Smith used a device to test the cable as it was being laid down. This test device used bars of selenium. Much to his surprise, he found that the resistance to the flow of electricity in the selenium varied in the lab. Sometimes the bars exhibited high resistance, sometimes they exhibited low resistance. Upon further investigation, Smith discovered that resistance changed when light was shone on the bars. Moreover, the change in resistance was proportional to the amount of light. The more light, the lower the resistance. Although Smith could not explain his discovery, today we realize that light aided the flow of electrons through the selenium bars. This puzzling phenomenon was later confirmed by researchers W. G. Adams and R. E. Day. They found that light shining on selenium actually created a tiny electrical current.

Smith's discovery led to research designed to find ways this phenomenon could be put to use. The first breakthrough came a few years later, when an American inventor, Charles Fritts, devised the very first electricity-producing solar cell. Fritts's solar cell consisted of a thin layer of selenium to which he applied an ultrathin transparent gold film. When exposed to light, this device produced a tiny electrical current. Although Fritts could reliably repeat his experiment, engineers were skeptical. How, they wondered, could this device produce electric energy without burning some type of fuel? Some physicists even claimed that Fritts's solar cell violated the laws of physics, specifically the laws of thermodynamics. The first law of thermodynamics states that energy can neither be created nor destroyed. With no apparent source of energy, many engineers viewed Fritts's device with skepticism.

Undaunted, Fritts sent his solar cells to Werner Siemens, a prominent German inventor and industrialist. Siemens tested the device and confirmed that it did indeed produce electricity when illuminated by sunlight. How it worked, Siemens could not say. But it did work.

As with many other discoveries of the time, it took scientists several decades to understand Fritts's simple solar cell. Today, thanks to the introduction of quantum mechanics, scientists know that light striking selenium atoms energizes their outer shell electrons. These energized electrons can be forced to flow through an external circuit. This effect is known as the *photovoltaic effect*. It's the basis of all solar cells.

Although Fritts's solar cells worked, they were extremely inefficient at converting sunlight to electricity. The scientist could never achieve efficiencies greater than 1%. Because of the low efficiency and the high cost of selenium, most scientists dismissed this technology as impractical. One exception, however, was the visionary German scientist Bruno Lange, who was also working on solar cells similar to Fritts's. In 1931, Lange prophesized that "in the not distant future, huge plants will employ thousands of these plates to transform sunlight into electric power...that can compete with hydroelectric and steam-driven generators in running factories and lighting homes." Unfortunately, Lange's solar cells were also terribly inefficient.

High costs and low efficiency impaired the development of solar electricity for many years, but in the early 1950s, research in solar electricity took an unexpected turn—a turn that would have profound consequences for this floundering technology. Two researchers, Calvin Fuller and Gordon Pearson, were experimenting with silicon rectifiers at Bell Labs. (A rectifier is a device that converts alternating current [AC] to direct current [DC] electricity.) Fuller and Pearson discovered that the efficiency of the rectifiers varied with the amount of light shining on them. Their studies also showed that sunlight striking their silicon-based rectifiers caused electrons to the flow through them—that is, it created an electrical current. Quite by accident, then, they had discovered another solar cell—this one, a silicon-based device that would revolutionize the industry. What is more, the efficiency of their solar cells was four times greater (4%) than that achieved by their predecessors, the selenium solar cells. Additional research and development performed by Fuller and Pearson with Daryl Chapin (who'd worked on selenium cells at Bell Laboratories) boosted the efficiency to 6%.

Excited by the prospects of their researchers' findings, Bell Laboratories began to explore ways to commercialize silicon solar cells. One application that the company hoped would prove lucrative was the use of PV cells to provide electricity to amplify long-distance phone signals at remote repeater stations. (A repeater station is an automated facility that greatly extends the range of communications by amplifying telephone signals. It consists of a receiver tuned to one frequency and a transmitter tuned to another, linked by a controller. When the receiver receives a weak signal, the controller activates the transmitter. It retransmits the signal at a higher strength. Repeaters are usually installed on top of tall buildings or on mountains.) Other companies pursued the use of PVs as well, for example, to power remote buoys for the Coast Guard and lookout towers for the US Forest Service. Unfortunately, the cost of the early PVs was too high, and the technology seemed doomed.

But then came satellites. To power satellites, the aerospace industry had concluded that it either had to install batteries or use some form of technology that could generate energy in space. To work, however, the batteries or energy technologies would have to be lightweight, compact, and reliable. Batteries failed on all three counts. Not only were they heavy and bulky, batteries had a limited lifespan. Solar cells, on the other hand, were relatively light and could generate a continuous supply of electrical energy from the intense solar energy found in outer space—and they could do it for decades. Who cared if they cost a lot? They were just what the doctors ordered—the Ph.D. type of doctors, that is.

Solar cells performed admirably in this application. As a matter of fact, since 1958 solar cells have powered virtually all satellites launched into space. Although early solar cells were extremely expensive, that didn't matter to NASA or the Department of Defense (DOD). There were no other safe, practical options.

The space industry owes a deep gratitude to solar cells; in turn, today's PVs owe their existence to the space program. However, the solar cells in the PV modules we install these days are not the same as those used to power satellites. Nevertheless, they are closely related cousins and are becoming an important source of electricity, helping to make the German solar scientist Bruno Lange's prediction come true.

What Is a PV Cell?

Photovoltaic cells are solid-state electronic devices like transistors, diodes, and other components of modern electronic equipment like computers and televisions. These devices are referred to as *solid-state* because electricity that controls and powers the various functions flows through silicon, a solid material that makes up computer chips that make these devices possible. Silicon chips have replaced vacuum tubes in older equipment. In a vacuum tube, electrons flow through a vacuum inside glass tubes.

Most solar cells in use today are made from one of the most abundant materials on the planet, silicon. To understand how a PV cell functions, though, you must understand a bit about the atomic structure and the behavior of electrons in silicon atoms. Before we explore these topics, though, let's take a brief look at light—the source of energy.

The Nature of Light

For years, scientists thought that solar radiation was emitted and transmitted as waves. However, studies revealed that light also exhibits properties of particles. Today, we know that light has a mysterious dualistic nature. That is, it behaves like waves *and* particles. This phenomenon, which has puzzled many a high school physics student over the years, is referred to as the *wave-particle duality*.

Light exhibits wave behavior when light waves interact with one another. During such times, light waves can cancel each other out if the troughs and peaks of the two waves are off by 180°. This phenomenon, known as *destructive interference*, is illustrated in Figure 3.1a.

Light also exhibits wavelike behavior when it interacts with matter of different density, for example, when light waves pass from air into a dense glass lens. This causes light waves to change speed and bend, a phenomenon known as *refraction* (Figure 3.1b).

However, light can also exhibit particle behavior. It does so when it interacts with matter. For example, when light strikes solid materials it transfers small amounts of energy to the atoms in these materials. That's why light shining on a

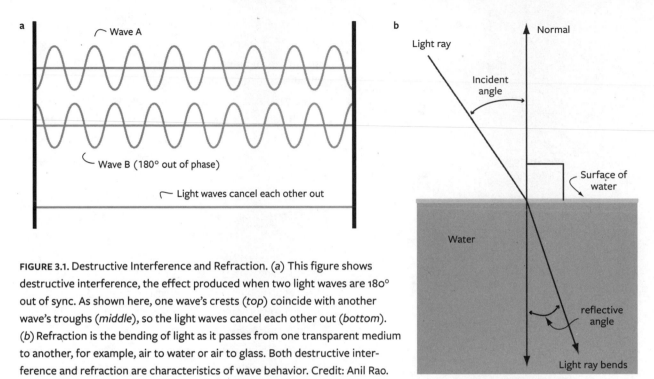

FIGURE 3.1. Destructive Interference and Refraction. (*a*) This figure shows destructive interference, the effect produced when two light waves are 180° out of sync. As shown here, one wave's crests (*top*) coincide with another wave's troughs (*middle*), so the light waves cancel each other out (*bottom*). (*b*) Refraction is the bending of light as it passes from one transparent medium to another, for example, air to water or air to glass. Both destructive interference and refraction are characteristics of wave behavior. Credit: Anil Rao.

dark object causes it to heat up. As you shall soon see, the energy light imparts in PV cells is sufficient to create an electric current.

Light striking PV cells is transmitted through space in discrete parcels, or packets of energy, called *photons*. A photon is considered an *elementary particle*, that is, a carrier of all types of electromagnetic radiation. That is to say, all forms of electromagnetic radiation from heat to ultraviolet radiation to light to X-rays consist of photons. Photons differ from other elementary particles, such as electrons, because they have no mass. As a result, photons can travel through a vacuum at the speed of light—186,000 miles per second (299,338 kilometers per second). Just to give you an idea how fast that is: A beam of light could travel around the equator (24,900 miles) almost seven and a half times in one second.

Conductors, Semiconductors, and Insulators

To understand how photons of light produce electricity in PV cells, let's begin by reviewing atomic structure. As most readers know, an atom is made up of a central region known as the *nucleus*. It is an extremely dense region containing positively charged particles called *protons* and uncharged particles called *neutrons* (Figure 3.2).

Atomic Structure

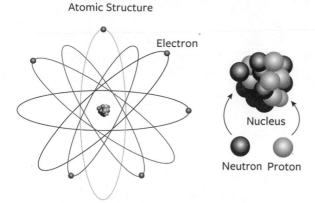

FIGURE 3.2. Atomic Structure. This drawing illustrates the structure of the typical atom. In all atoms, electrons orbit around a central nucleus containing the protons and neutrons. Credit: Forrest Chiras.

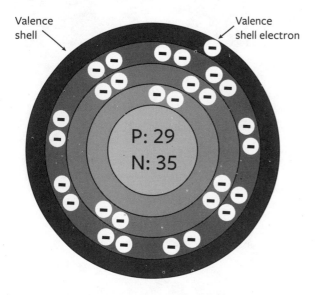

FIGURE 3.3. Electron Shells. The electrons of an atom occupy different shells based on their energy. The most energetic electrons are located farther away from the nucleus. The electrons in the outermost shell (known as the valence shell) are responsible for the chemical and physical properties of atoms. Credit: Forrest Chiras.

Orbiting around the nucleus are extremely low mass, negatively charged particles known as *electrons*. They occupy a region called the *electron cloud*. It constitutes the bulk of the atom's volume.

Electrons live in different regions of the electron cloud. These are known as *electron shells*, as illustrated in Figure 3.3. Their position in the electron cloud is determined by their energy. The most highly energetic electrons are found in orbits further from the nucleus.

In electrical terms, elements can be classified in three broad categories based on electrical properties: conductors, insulators, and semiconductors. What determines their classification?

The answer is the number of electrons in the outermost region (shell) of the electron cloud of the atoms. This region is known as the *valence shell*, and the electrons are called *valence shell electrons*.

Valence shell electrons are responsible for the electrical as well as many of the chemical properties of atoms. For example, the number of electrons in the outer shell determines how atoms chemically bond to one another to form molecules. To understand how the number of electrons in the valence shell affects electrical properties, let's start with conductors.

Conductors are atoms that have only one or two electrons in their outermost shell (Figure 3.3). These electrons can be easily ejected—that is, forced out of their orbit around the nuclei of their atoms. When propelled out of the valence shell, these electrons are said to be pushed into the *conduction band*. The conduction band is a range of electron energy that's higher than the valence shell. Electrons possessing this energy are free to move about and, more important, to accelerate under the influence of an applied electric field. Put more simply, they can be forced to move if voltage is present. You will recall from Chapter 1 that voltage is an electromotive force. It pushes electrons along conductors. When voltage moves electrons along a conductor, an electric current is created. Electric currents, then, consist of excited (energetic) electrons that flow from one atom of a conductor such as copper to the next along the length of the conductor.

Insulators, on the other hand, are atoms that have six to seven electrons in their outer valence shells. These electrons cannot be easily jolted out of the atom and made to flow. Thus, materials made from such atoms resist the flow of electricity. Insulators are used to coat wire conductors made from copper or aluminum. The coating contains the current within the conductor and prevents shock (even lethal electrocution) and fires. That's why it is so important to protect the insulation coating on wires from being abraded or cut when installing a PV system—or any electrical wiring for that matter.

Semiconductors occupy a middle ground. They have three to five electrons in their valence shells. They are marginally conductive. Under certain circumstances, these electrons can be "persuaded" to leave their orbit, to enter a higher energy state called the *semiconductor band*. This allows them to flow from one atom to another, creating tiny electrical currents.

Silicon atoms are the main component of PV cells. These atoms have four outer valence shell electrons. What boosts these electrons out of their valence shells in a PV cell?

Light.

Photons with sufficient energy can eject electrons from the valence shell of silicon atoms, forcing them into the semiconductor band. When an electron leaves a silicon atom, it creates an empty space known as a *hole* in the atom from which it came. This hole can be filled by an electron ejected from a neighboring atom. Sunlight striking silicon therefore creates a wild session of musical chairs, although no chairs are withdrawn. Holes are continuously created and filled.

This wild movement of electrons does not result in an orderly flow of electrical current. In fact, the electrons knocked loose by photons drift around aimlessly until they find a vacant hole. When that happens, the energy of the photons is converted to heat, not useful electricity.

To produce an electrical current, PV cells need something that prevents electrons from falling into holes. This is achieved several ways.

Scientists have found that they can turn this pointless game of musical chairs into electrical current—that is force electrons to the surface of a solar cell—by chemically modifying silicon, specifically adding two more elements, boron and phosphorus atoms, in small quantity to silicon (Figure 3.4). This process, called *doping*, cleverly forces electrons to flow in only one direction, creating an orderly flow of electrons out of the solar cell. How is this achieved?

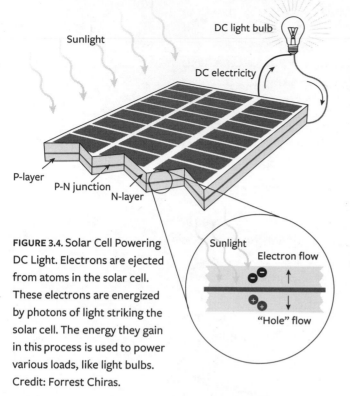

FIGURE 3.4. Solar Cell Powering DC Light. Electrons are ejected from atoms in the solar cell. These electrons are energized by photons of light striking the solar cell. The energy they gain in this process is used to power various loads, like light bulbs. Credit: Forrest Chiras.

How Solar Cells Work

Most solar cells in use today are thin wafers of silicon about $\frac{1}{100}$ of an inch thick (they range from 180 μm to 350 μm in thickness). As shown in Figure 3.5, most solar cells consist of two layers—a very thin upper layer and a much thicker lower layer. The upper layer consists of silicon doped with phosphorus atoms; the bottom layer contains silicon doped with boron atoms.

As illustrated in Figure 3.6, silicon atoms in solar cells have four electrons in their outermost shells. Each of these electrons is "shared" with a neighboring silicon atom to form highly ordered silicon crystals. The upper layer of a PV cell, known as the *n-layer* (n stands for negative) contains phosphorus atoms. Each phosphorus atom has five electrons in its valence shell. Physicists believe that the presence of phosphorous atoms introduces surplus electrons in the n-layer. These rogue electrons can be propelled into the conduction band when struck

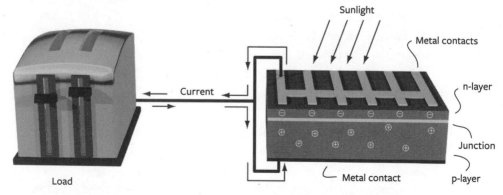

FIGURE 3.5. Diagram of Solar Cell Showing N-Layer and P-Layer. Solar cells like the one shown here consist of two layers of photosensitive silicon, a thin top layer (the n-layer), and a thicker bottom layer (the p-layer). Sunlight causes electrons to flow from the cell through metallic contacts on the surface of most solar cells, creating DC electricity. Solar-energized electrons then flow to loads where the solar energy they carry is used to power the loads. The electrons then flow back to the solar cell.

by photons containing a sufficient amount of energy. Here they are free to move about.

As shown in Figure 3.6, the lower layer of the PV cell is known as the *p-layer*. (P stands for positive.) It is doped with boron atoms. Boron contains three electrons in its outermost (valence) shells. This, in turn, creates "holes" in the crystalline structure of the p-layer. These holes can be filled by free electrons—electrons ejected from neighboring atoms. Because holes created in one atom can be filled with electrons from another atom, the holes appear to move about.

So how does a PV cell work?

The simple answer is that sunlight energy causes electrons to be released from silicon atoms in solar cells. Doping forces electrons to the surface, that is, the n-layer. These solar-energized electrons are then gathered by thin metal contacts (hair-thin silver wires) located on the face of solar cells. The electrons are picked up by the silver contacts and flow together to create an electrical current. It is drawn from each solar module by wires on the back, called *leads*. Electricity from all the solar modules is "pooled" to create a high-voltage electric current. That's the simple answer.

For those who'd like a scientific explanation, the story is as follows: When the n-layer of a PV cell is formed, an electric field is created instantaneously in the cell. (It has a voltage of about 0.6 volts.) This tiny electrical field is created by the flow of electrons from the n-layer to the p-layer when the n-layer is first created. This, in turn, creates a charge imbalance at the junction of these two layers, the n- and p-layers. The top part of the junction is positively charged (because electrons have left). The lower part is negatively charged (because they've set up residence there).

The positively charged regions of the electrical field, physicists hypothesize, draws electrons liberated

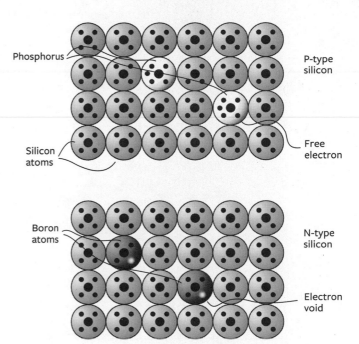

FIGURE 3.6. Atomic Structure of N- and P-Layers. This drawing shows the crystalline structure of both the n-type and p-type layers of a solar cell. As shown here, pure crystalline silicon atoms have four electrons in their outermost shell. Each of these electrons bonds with a neighboring silicon atom to form silicon crystals. Because the four electrons bond to neighboring atoms, they are not free to move. Phosphorus atoms in the upper layer (the n-layer) have five electrons in their outmost shell. The surplus electrons in phosphorus atoms in this layer are free to move about when struck by photons containing enough energy to eject the electron. Boron is present in the bottom layer. Boron contains three electrons in its outermost shell and thus creates holes in the crystalline structure of the boron-doped p-layer that can be filled by free electrons—that is, electrons ejected from neighboring atoms. Credit: Anil Rao.

in the deeper p-layer into the superficial n-layer when sun penetrates a PV cell. The negatively charged portion of the electrical field also stops electrons liberated in the n-layer from flowing into the p-layer when a cell is illuminated by sunlight. (The negatively charged lower part of the junction repels electrons.) The net effect is that electrons can only move in one direction—from the p- and n-layers to the surface of the cells where the contacts are located. As noted in Chapter 1, numerous solar cells are wired in series in a solar module. Because of this, electrons extracted from one cell flow to the next cell, and then to the next cell, etc., until they reach the negative lead of the module. After leaving the PV module, electrons flow through an external circuit to a load (a load is any device that consumes electricity). After delivering the energy they gained from sunlight to the load, the electrons return to the positive lead of the module. The de-energized electrons then flow back into the solar cells from whence they came. They fill the holes in the p-layer and permit the circuit to continue *ad infinitum*. For an even more detailed description, see the sidebar, "How PVs Work."

FIGURE 3.7. How PV Cells Work. These three drawings illustrate how an electrical field is created at the P–N junction of the solar cell. (*a*) The formation of the n-layer, resulting from the infusion of phosphorous ions, creates a surplus of electrons. Immediately after the formation of the n-layer, these electrons migrate to holes in the underlying p-layer. (*b*) This results in a charge separation. The charge separation, in turn, creates an electric field. (*c*) The electric field is responsible for the movement of photo-excited electrons from the p-layer into the n-layer, where they are picked up by the silver contacts. Credit: Forrest Chiras.

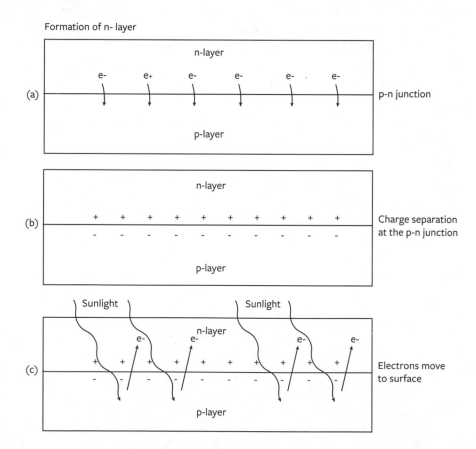

How PVs Work

PV cells are relatively simple devices governed by complex physical and chemical interactions. To understand how a PV cell works, we begin with the formation of the top layer, known as the n-layer. It is created by injecting phosphorus atoms (described later) into the superficial layer of thin silicon wafers, the precursors of solar cells.

Immediately after the n-layer is formed, some of the excess electrons from the phosphorus atoms in this layer diffuse across the junction between the n-layer and the underlying p-layer, as shown in Figure 3.7a. These electrons fill the holes in the neighboring p-layer created by the addition of boron atoms during the production of PV cells. The flow of electrons from the n-layer to the p-layer results in a depletion of electrons in the n-layer near the p-n junction. This, in turn, results in an increase of electron holes in the n-layer at p-n junction. An atom with an electron hole is positively charged ion.

The flow of electrons into the p-layer near the junction, however, results in a surplus of electrons on that side of the junction. The electrons fill up holes in boron atoms located in that region. The addition of electrons gives this region a net surplus of negatively charged particles and an overall negative charge. As just noted, the separation of positive and negative charge creates an electric field across the junction— that is, it creates voltage, a mysterious force that pulls electrons toward the surface of a PV cell.

When a photon of light ejects an electron into the conduction band in the p-layer and that electron drifts near the junction, it is attracted by the electric field and swept across the junction to the n-layer (Figure 3.7c). Because there are more electrons on the p-side than the n-side, electrons cannot travel in reverse.

The negatively charged surface of the n-layer at the p-n junction blocks electrons from moving back from whence they came, the p-layer. The same applies to electrons liberated by solar energy in the n-layer. These electrons cannot flow backward, into the p-layer or into the depths of a solar cell. They are blocked by the junction. (Physicists refer to this type of junction as a junction diode. Junction diodes allow current to flow in only one direction.) Electrons that accumulate in the p-layer are carried away by hair-fine threads on the front surface of solar cells.

As electrons move across the junction, from the deeper parts of the p-layer, electron holes are created in the p-layer. These are filled by electrons liberated deeper in the solar cell by sunlight. Eventually, however, all holes are filled by de-energized electrons flowing into the back side of the PV cells. This occurs because there is an ultrathin layer of conductive metal on the back of each solar cell. It feeds electrons into the solar cell, permitting the cell to continue to function.

When light shines on and into a PV cell, electrons flow out of the cell from the n- and p-layers, through the front metal contacts, through a wire (lead) connected to each module. The leads of one module are connected to the next. They eventually terminate on a set of wires that transports the DC electricity produced by the array to an inverter, which converts DC into AC electricity (the type of electricity we use in homes and businesses). After their energy is stripped from them in the inverter, the once solar-energized electrons return to the array by a different wire. It delivers the de-energized electrons to the solar cells of the modules.

Types of PV

Solar cells can be made from a variety of semiconductor materials. By far the most common is silicon. As noted earlier, pure silicon is extracted from silicon dioxide, derived from two sources: quartzite and silica sand. Quartzite is a rock made entirely of the mineral quartz. It contains nearly pure silica (silicon dioxide). Silica sand is clean white sand containing a high percentage of silica (silicon dioxide). Silica sand is derived from quartz.

Although not the most efficient in converting sunlight to electricity, silicon dominates the semiconductor market. That's because silicon semiconductors produce the most electricity at the lowest cost.

Three forms of silicon are used to make solar modules: monocrystalline, polycrystalline, and amorphous thin-film. Monocrystalline and polycrystalline are considered first generation solar electric technology. Amorphous silicon (thin-film) is a newer innovation, and is considered second-generation PV technology.

Monocrystalline PVs

Monocrystalline cells—a.k.a. *single-crystal cells*—were the very first commercially manufactured solar cells. They've been around since the mid-1950s. Monocrystalline cells are made from ultrathin wafers sliced from a single crystal of pure silicon known as a *silicon ingot* (Figure 3.8). The long, cylindrical single-crystal ingots (a.k.a. *column ingots*) are made by melting highly purified chunks of silicon and a trace amount of boron. Once melted, a seed crystal is dipped in the molten mass (also known as the *melt*) of silicon and boron. The seed crystal is rotated and slowly withdrawn. As it's withdrawn, silicon atoms from the melt attach to the seed crystal, exactly duplicating its crystal structure. As the crystal is slowly withdrawn, it grows larger and larger. Eventually, the ingot may grow to a length of 40 inches (100 cm) and a diameter of 6 to 8 inches (15 to 20 cm).

Once fully extracted from the melt, the ingot is cooled. In the old days, for example, in the 1970s and 1980s, the round ingot was sliced into thin, round wafers that were assembled into a module. However, manufacturers could not "pack" many cells into a module, and there was a lot of wasted space. To avoid this problem, manufacturers trim

FIGURE 3.8. Monocrystalline PV Cells. This ingot is a huge crystal of silicon that is sliced to make monocrystalline PV cells. Credit: SolarWorld.

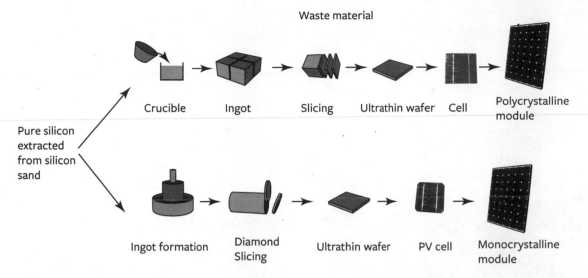

FIGURE 3.9. Steps in Wafer Production. As shown here, silica is used to produce high-purity silicon, which is then used to make monocrystalline and polycrystalline solar electric cells. Please note that waste from the production of monocrystalline solar cells is also used in the production of polycrystalline solar cells. Credit: Forrest Chiras.

the round ingot to produce a nearly square ingot—a square ingot with rounded corners. It is then sliced with a diamond wire saw, producing ultrathin wafers (Figure 3.9). Trimming an ingot prior to slicing allows manufactures to place more solar cells in a module, which increases the output per square cm. (You can fit a lot more nearly square solar cells in a module than round cells.) Waste from this process is remelted and reused often to make the next type of solar cell, polycrystalline.

Because monocrystalline ingots are "squared off," modules made from them have small white spaces (interstices). They are easily recognizable from a distance, as shown in Figure 3.10.

Monocrystalline cells boast the highest efficiency of all conventional PV cells, with commercially available cells attaining something in the range of 16% to 18%. Efficiency is a measure of energy input vs. energy output. Generally speaking, if a

FIGURE 3.10. Monocrystalline Module. Monocrystalline solar cells are sliced from a long cylindrical ingot of silicon doped with boron. You can identify the type of solar cell by its distinct shape, as described in the text. Credit: Dan Chiras.

module is 16% efficient, that means that for every 100 units of sunlight energy, the module will produce 16 units of electrical energy. Although you may occasionally read about even-higher efficiencies, these solar cells tend to be still in development and way too costly for commercial production.

Polycrystalline PV Cells

Polycrystalline solar cells are a relative newcomer. They emerged on the scene in 1981. Just like monos, they are made from pure silicon with a trace of boron. To make a polycrystalline cell, however, the molten material is poured into a square mold. It is then allowed to cool very slowly. As the ingot cools, many smaller crystals form internally. Once cooled, the cast ingot is removed from the mold, then sliced using a diamond wire saw, creating wafers used to fabricate solar cells.

Polycrystalline solar cells in previous years, for example, the mid-1990s, contained many larger crystals, like the ones shown in Figure 3.11a. Today, however, silicon crystals in these modules are often so small that they are undetectable to the naked eye (Figure 3.11b).

Polycrystalline solar cells are slightly less efficient than monos—on average, they are about 15% to 17% efficient. The reason for this is that the boundaries between these crystals in a solar cell resist the flow of electrons through the crystal, reducing the output of a cell. Physicists believe that these boundaries also create opportunities for electrons to fall back into holes in the silicon crystal from whence they came before they can be whisked away by the surface contacts. So why bother making them?

Although polycrystalline cells are slightly less efficient, they require less energy to produce. Because of this, they're cheaper to manufacture. Manufacturers, in turn, have passed the savings onto customers. In my experience, polys are typically about 15% cheaper than monos.

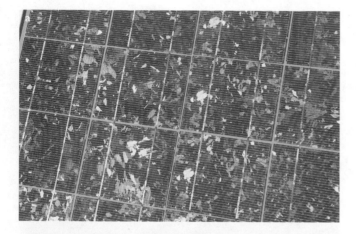

FIGURE 3.11. (a) Close-up of Older Polycrystalline Cell Showing Crystals. (b) Newer Polycrystalline Cell. Older polycrystalline solar cells contained very large crystals. Newer polycrystalline solar cells contain very small crystals which are extremely difficult to see. Credit: Dan Chiras.

Modules made from polycrystalline cells have a distinctive appearance. Unlike modules manufactured from monocrystalline cells, there's virtually no wasted space. The square or rectangular cells can be tightly packed into a module. (You won't see any open space between solar cells.)

Although polycrystalline modules are slightly less efficient, it is important to note that tight packing makes up for the cells' lower efficiency. As a result of the higher packing density and slightly larger module size, a 285-watt module made from polycrystalline PV cells will produce the same amount of electricity as a 285-watt module made from monocrystalline PV cells. Don't assume that you are better off buying modules with slightly more efficient cells.

According to some sources, polycrystalline modules also perform a little better under high temperatures. Because of these factors, I advise my students and clients to avoid the must-have-the-most efficient—that is, monocrystalline—module mentality. Check out polycrystalline modules—they could save you a substantial amount of money.

Module vs. Cell Efficiency

It is important to note that although a manufacturer may install a 16% to 18% efficient solar cell, the efficiency of the module will be around 2% lower. The reason for this is that in a PV module, the cells are connected by thin strips of solder. Electric connections and conductors in a solar cell create a tiny amount of resistance to the flow of electricity, lowering the efficiency of a solar module.

Ribbon Polycrystalline PV

While most polycrystalline cells are sliced from ingots, wafers can also be made in long continuous ribbons. Although a dozen different technologies have been developed to make ribbons, only four have been commonly used. In one of these techniques, shown in Figures 3.12a and b, a seed crystal is attached to two heat-resistant wires called *filaments*. The filaments are immersed in a molten mass of silicon and then slowly drawn from the melt. The ribbon grows linearly as the molten silicon that spans the wires solidifies. Once the ribbon is completely formed, it can be sliced into rectangular wafers. Figure 3.13 shows another, similar process used by the now-defunct Evergreen Solar, a US company. As you can see, they produced two ribbons simultaneously from each furnace.

Although the efficiency of ribbon silicon wafers is slightly lower than other PV cells, the technology offers several advantages. One of them is that its production requires considerably less energy than monocrystalline and polycrystalline production. The ribbon technology also eliminates a lot of time-consuming, wasteful, and costly ingot slicing.

Although ribbon polycrystalline modules have many advantages, I can't find many of the modules on the market today. Solartech is one of the few companies that sells them, but it appears as if the company is no longer producing them.

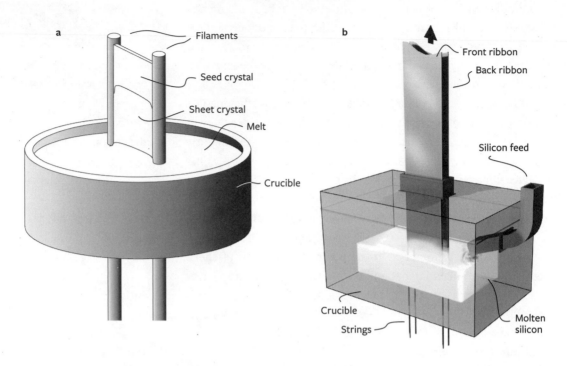

FIGURE 3.12. Formation of Ribbons. (*a*) This drawing shows one of several techniques used by PV manufacturers to produce ribbons of PV polycrystalline material. They are then sliced to create solar cells mounted in modules. (*b*) This drawing shows the ribbon making technology employed by Evergreen Solar. Credit: Anil Rao.

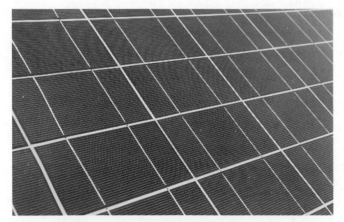

FIGURE 3.13. Ribbon Technology Used by Evergreen Solar. Solar cells in an Evergreen Solar module. Note that the solar cells are elongated rectangles.

Adding the N-layer and Building the PV Cell

All three of the processes described so far result in the formation of the p-layer, the thicker, bottom layer of PV cells. How is the much thinner, phosphorus-containing n-layer added?

After the p-layer has been created, the wafer is dipped in a solution of sodium hydroxide. This powerful solution removes impurities deposited on the wafers during manufacturing. It also etches (roughens) the surfaces of the wafer. Etching allows subsequent coats to adhere better. It also reduces reflection, improving the efficiency of solar cells. (Reduced reflection means more sunlight is absorbed by the cell.)

Once etched, the wafers are placed in a diffusion furnace and heated. Phosphorous gas is introduced into the furnace. In this high-temperature environment, phosphorus atoms are shot into (penetrate) the exposed top surface and sides of the wafer. (The bottom side is masked to protect it from the gas.) This results in the formation of a very thin, but complete n-type layer (Figure 3.14).

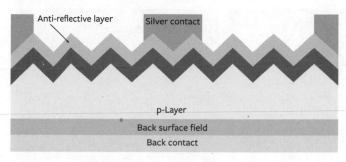

FIGURE 3.14. Cross Section through PV. This drawing shows details of the many layers of a traditional mono and poly silicon PV cells. Credit: Forrest Chiras.

After a wafer is removed from the diffusion furnace, its edges are shaved off—very delicately, of course—to keep the n-layers from contacting the underlying p-layer along the perimeter of the cells.

Metal contacts are then screen-printed onto the face (the n-layer side) of the wafer using a silver paste. The silver paste is then baked on. Contacts are applied in a grid pattern consisting of two or more wide main strips and numerous ultrafine (hair-thin) strips that attach perpendicularly to the main strips (Figure 3.14). This grid collects the electrons that gather in the n-layer, and forms the negative connection to the cell. Because silicon is highly reflective, an anti-reflective coating (ARC) of silicon nitride or titanium oxide is applied to the surface of the cells. This gives the solar cell a much darker color (remember, silicon is milky white.) The black or blue-black ARC reduces the reflection of light off solar cells by 35%, resulting in much greater absorption of the Sun's energy and greater output.

As shown in Figure 3.14, another layer is added to the cell. It is called the *back surface field*. The back surface field is yet another innovation designed to improve the efficiency of solar cells. It does so by reducing the chances of electrons ejected out of semiconductor material from falling back into vacant holes. It consists of strips of conductors.

Once completed, the cells are individually tested for voltage and current output under controlled conditions using artificial light. There's a slight variation in voltage among cells due to several factors beyond the scope of this book. The cells are then sorted according to their voltage and assembled using cells with similar voltage. (That's why a manufacturer will produce 275-watt modules, 280-watt modules, and perhaps even 285-watt modules. They're from matching cells.)

After being sorted, solar cells are assembled in a module and robotically soldered, creating series strings that boost the voltage. The cells are then encapsulated in a thin sheet of clear plastic, known as EVA (ethylene vinyl acetate). Glass is applied to front of the module and another thin layer of plastic (usually white)

FIGURE 3.15. The Grid on a PV Cell. This close-up photo of poly-crystalline solar cell shows the individual crystals that form when molten silicon is cooled in the mold. Credit: Dan Chiras.

is attached to the back, creating a watertight seal. In some modules, glass is substituted for the back plastic. This creates a more translucent module that allows light to pass through, which is ideal for the roofs of car ports, patio roofs, bus stops, and other outdoor structures where partial shading is preferable.

Virtually all solar modules on the market today come with aluminum frames. These frames are anodized to prevent oxidation. Frames are used to attach modules to racks. They also become part of the equipment, grounding the PV array by providing a pathway for electricity to safely flow from a damaged wire, for instance, to the ground to prevent shocks.

Thin-film Technology:
Second-generation Solar Electricity

In an effort to produce solar modules at a lower cost—which means using less energy and less material—several manufacturers have turned to a new technology, known as *thin-film*. Thin-film modules are the second generation of solar electricity.

Unlike manufacturers of previous technologies, thin-film producers manufacture entire modules rather than individual cells that are later assembled into modules. Skipping the step of assembling cells into modules saves time, energy, and money.

In silicon-based thin-film solar designs, known as *amorphous silicon*, silicon is deposited directly onto a metal backing (aluminum), glass, or plastic by a technique known as *chemical vapor deposition*. This creates a thin film of photo-reactive material. Once it has solidified, a laser is used to delineate cells and create connections between the newly formed cells.

Thin-film PVs are primarily made from silicon. These are known as *amorphous thin-film* to distinguish it from crystalline silicon cut from ingots. Three other semiconductor materials are also used to make thin-film solar modules: (1) cadmium telluride (CdTe), (2) copper indium gallium diselenide (CIGS), and (3) gallium arsenide (GaAs). The vast majority of the thin-film solar on the market today is CdTe (50%), CIGS (25%), or amorphous silicon (25%).

One of the advantages of thin-film is that it can be manufactured in long, continuous rolls, or incorporated onto flexible substrates, creating PV laminates that can be applied to standing seam metal roofs.

Short pieces of laminated PV materials can be sewn onto backpacks and gym bags and used to charge cell phones and other portable electronic devices. Thin-film is also used by the military to power portable communication devices. Some thin-film products can even be applied to plate glass windows and skylights to generate electricity. (Like solar roof tiles, this is a form of *building-integrated PV*.)

The vast majority of thin-film solar, however, is sandwiched between two layers of glass to create modules—albeit rather heavy ones. (They are about twice as heavy as standard polys and monos.) Modules made from CdTe are currently used

Energy Payback for PVs

You may have heard people say that it takes more energy to make a PV system than you get out of it over its lifetime. Fortunately, that's not even close to being accurate.

While it takes energy to make solar cells, modules, and the remaining components of a PV system, the energy payback is amazingly short—only one to two years, according to a study released in 2006 by CrystalClear, a research and development project on advanced industrial crystalline silicon PV technology funded by a consortium of European PV manufacturers. As Justine Sanchez notes in her 2008 article in *Home Power*, "PV Energy Payback" (Issue 127): "Given that a PV system will continue to produce electricity for 30 years or more, a PV system's lifetime production will far exceed the energy it took to produce it."

CrystalClear's research showed that it takes two years for a PV system with monocrystalline solar cells to make as much energy as was required to manufacture the *entire* PV system. The researchers also calculated that it took 1.7 years for a polycrystalline PV system and 1.5 years for modules made from ribbon polycrystalline PVs to produce as much energy as was required to make them. According to the National Renewable Energy Laboratory, most thin-film modules, which require even less energy to produce, achieve energy payback in one year. Some have even shorter paybacks. Thin-film modules made from cadmium telluride, for instance, have an energy payback of eight months in ideal conditions.

These energy payback studies were performed for sunlight conditions similar to those found in southern Europe, which has an average insolation of 4.7 peak sun hours. That's the same for eastern Missouri and western Illinois. For those who live in sunnier climates, the energy payback would be even quicker. For those who live in less-sunny regions, the energy payback would be slower.

Most of the energy required to make a PV system goes into producing the modules—about 93% of the entire energy budget.

As just noted, the most energy-intensive modules are those made from monocrystalline solar cells. Polycrystalline cell modules require 15% less energy to manufacture than monocrystalline modules. Ribbon cell module production requires 25% less energy than monocrystalline and about 12% less than polycrystalline. Thin-film uses even less energy, about 50% less than monocrystalline modules.

FIGURE 3.16. Solar Calculator and Solar Watch. The author's solar calculators and his Citizen solar watch use thin-film PV to produce electricity from sunlight as well as room light to charge a small battery. As a result, their batteries never need replacement. Credit: Dan Chiras.

in several large commercial solar farms in Australia and the United States. Two commercial arrays are located in the desert of southern California. Each array has a rated capacity of 550 megawatts—that's the size of a large coal-fired power plant. They provide electricity to utilities in California, one of the world's leaders in renewable energy.

Thin-film got its start as the power source for calculators, watches, and similar devices (Figure 3.16). Although they performed well under artificial light, the cells used for these technologies were only about 4% to 5% efficient. The also degraded in intense sunlight, rendering them useless for solar modules.

To address these issues, thin-film PV manufacturers applied several layers of amorphous silicon to their modules. This increases their efficiency—to around 8% to 9%—and prevented photodegradation. Even so, they were only about half as efficient as crystalline solar.

Today, thanks to improvements (discussed shortly), at least one manufacturer (Stion) produces a 14% efficient thin-film. There's even a 16% thin-film module, but it's only for large commercial arrays.

To improve efficiency, manufacturers apply two or more layers of *different* semiconductor materials. These modules absorb nearly the entire solar spectrum, including visible light, some infrared, and some ultraviolet (UV) radiation.

Multiple-layer cells are called *multi-junction cells*. They are the most efficient cells produced in the world. Two- and three-junction cells boast efficiencies of slightly over 40% and slightly more than 33%, respectively. Bear in mind, however, that these are research cells—the products of research laboratories—that are too expensive to manufacture for most applications. Because they are costly, high-efficiency multi-junction modules are used in outer space to power satellites. They are also being used to produce highly efficient concentrated PVs. These are PV arrays with lenses that concentrate light on the PV material (discussed shortly).

Thin-film technology offers several advantages over single and polycrystalline solar cells. One of the most important is that it requires considerably less energy to produce. Less energy is required because their manufacture eliminates the costly and energy-intensive ingot production and wafer slicing required to produce mono- and polycrystalline PV cells.

Another advantage of thin-film solar is that it uses considerably less semiconductor material. However, some materials used in thin-film modules are prohibitively costly. For example, tellurium used in CdTe thin-film modules is as rare as platinum, and as just as pricey. Even silicon, which is abundant, is energy intensive to purify and, hence, expensive.

One advantage of amorphous thin-film PV is that it is less sensitive to high temperatures. At 100°F, a crystalline module will experience a 6% loss in production, while a thin-film amorphous silicon module will experience only a 2% loss. This makes thin-film a good choice for hot, sunny climates like Florida and the tropics.

Thin-film is also more shade tolerant than crystalline PV. As Erika Weliczko points out in an article in *Home Power* magazine, "many thin-film cells are as long as the module itself, so shading an entire cell is more difficult than traditional 5- or 6-inch square or round crystalline cells." Thin-film modules also often incorporate bypass diodes that allow current to flow around or bypass shaded cells. (Though this is nothing new; crystalline solar modules also have bypass modules to lessen the effect of shading). Why worry about shading? Shading reduces the amount of sunlight falling on a module, and it also increases electrical resistance inside PV cells. Resistance, in turn, reduces the flow of electricity through a module, decreasing its output.

For years, the chief disadvantage of thin-film PVs has been their low efficiency. For many years, most thin-film products on the market had efficiencies in the 6% to 9% range. An array of thin-film modules, therefore, required about twice as much roof space as an array of crystalline silicon modules. Today, the efficiency of thin-film modules for residential installations has increased, so this is no longer a concern.

Another potential downside of thin-film modules is that it takes six months for them to reach their stable, rated output. What that means is that thin-film modules initially produce higher-voltage electricity. Crystalline modules stabilize immediately. Because the output of thin-film is 10% to 30% higher at first, care must be taken when sizing components of a PV system like inverters so the array doesn't exceed their input voltage range. The initial high-voltage output therefore makes it more challenging to design a system.

Another key disadvantage of thin-film PVs is that the modules produce high-voltage, low-current electricity. The higher the voltage, the fewer the number of modules that can be placed in series. (The National Electrical Code limits voltages in residential grid-tied systems to no more than 600 volts.)

Thin-film modules represent a tiny portion of total PV production and probably won't be an option for most readers.

Rating PV Modules and Sizing PV Systems

PV manufacturers list various parameters of their products on promotional material and spec sheets to assist buyers. These include rated power, power tolerance, and efficiency. Manufacturers also include several additional parameters, including short-circuit current, open-circuit voltage, maximum power voltage, maximum power current, maximum system voltage, and series fuse rating. (These are listed on a label on the back of PV modules.) This last group is used to compare PV modules and to design PV systems to ensure that wires (conductors) and various components can handle the voltage and current produced by an array under all conditions.

Rated power is the wattage a module produces under standard test conditions (STC). It is used to compare one module to another, for example, to compare price. As just noted, rated power is also useful when sizing a PV array. Installers, for instance, first determine electrical demand of a home, then determine the size of the system needed to produce it. Residents of a home, for example, might consume 12,000 kilowatt-hours a year. In eastern Missouri, that would require a 9,000-watt or 9-kW array. If you were installing 300-watt modules, you'd need 30 modules.

The next measurement, *power tolerance*, is the range within which a module either overperforms or underperforms its rated power at STC. It may be expressed as a percentage or as watts. Rated power tolerance ranges from –5% to +5%, depending on the manufacturer, although most modules today boast rated tolerances of –0/+5%. A 300-watt module with a power tolerance of –0/+5 watts is guaranteed by the manufacturer to produce 300 to 315 watts under STC.

Some manufacturers also list *rated power (watts) per square foot* of module. As the name implies, it is the power output per square foot of module area. This number is useful to those who have a limited amount of space to mount a PV array and want to generate as much electricity as possible in that space.

Module efficiency is also a handy number. As you have learned, efficiency is the ratio of output power (watts/m²) from a module to input power (watts/m²) from the Sun. An 18% module efficiency means that 18% of the incident (incoming) solar radiation is converted into electricity. Like rated power per square foot, it's handy for those who have limited space.

Five more parameters are also of importance. The first two are open-circuit voltage (V_{oc}) and short-circuit current (I_{sc}). They are important when it comes to sizing electronic components of a PV system, like inverters and conductors (wires). To understand them, let's begin with the graph shown Figure 3.17, known as a *I-V curve*.

The *current-voltage*, or *I-V*, curve shows the relationship between the current (I) and voltage (V) produced by any PV device—a cell, module, or array—under

PV Degradation

Thin-film and crystalline PV modules degrade over time, approximately 0.5% to 1.0% per year. According to some sources, degradation occurs because of chemical reactions between boron and oxygen with exposure to sunlight. SunPower, a major US manufacturer of solar modules, claims that degradation occurs as a result of cracking of silicon due to expansion and contraction caused by exposure to hot and cold temperatures. Corrosion of the silver metal contacts may also result in deterioration over time.

Whatever the cause, manufacturers provide a production warranty with their products (in addition to a materials warranty). Warranties are fairly standard in the industry. Every manufacturer I know of guarantees that their modules will still be producing 90% of their minimum peak power in ten years and 80% in 25 years. What this means is that if a 100-watt module has a –0/+5% power tolerance rating, in 10 years, the company guarantees it will produce 90 watts under STC in 10 years and 80 watts in 25 years under STC.

a specific set of conditions (cell temperature and irradiance). V_{oc} and I_{sc} are key points on the I-V curve.

One of the most useful parameters listed on modules and module spec sheets is the short-circuit current. *Short-circuit current* (I_{sc}) is the current that flows from an array if the output terminals (the positive and negative leads) are connected to one another, that is, shorted. While short-circuiting an electrical device like a battery can be very dangerous and destructive, short-circuiting a PV module is not. PV devices are *current limited*, meaning they can only produce a certain amount of current under optimal conditions—and the module is designed so that that amount of current will not damage it. That is, the internal wiring is robust enough to be able to conduct the current without overheating and causing damage.

The short-circuit current of an array is used for wire sizing in a PV system. More specifically, it is used to calculate the amount of current (amperage) that will flow through various wires to the inverter and hence the size of wires that need to be installed to conduct that current safely. Once he or she knows the wire size and the amount of current it can safely conduct (known as its *ampacity*), an installer can select the fuses or circuit breakers that are required by Code. Fuses and circuit breakers need to be installed to protect the wires and array

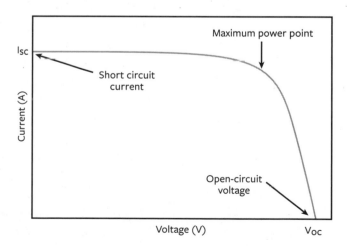

FIGURE 3.17. I-V Curve. Credit: Anil Rao.

from excessive current that could be introduced by a faulty inverter or charge controller, or from other abnormal conditions. That is, fuses and circuit breakers are designed to protect the wires against unusually high amperage. If a current in a wire exceeds a circuit breaker's rating, which is the wire's ampacity, a breaker will trip (open the circuit). This prevents current from flowing through the wire until the problem can be fixed. Excess current, that is, current that exceeds a wire's ampacity causes overheating that can melt the protective layer of insulation. This can cause sparks that lead to fires or expose conductors inside wires that could then cause shock or electrocution (fatal shock).

Short-circuit current is the maximum current a module can produce under most conditions. The only time a module will produce higher current is when the irradiance exceeds 1,000 watts/m^2. This is rare in most locations but can occur in high elevations like in mountain communities in the Rockies, Sierra Nevada, or Cascade mountains. In these locations, irradiation can climb as high as 1,250 watts/m^2. As a safety factor, then, the National Electric Code requires all DC wires in a PV system be rated for 25% higher current.

Another extremely important measurement that's used to design PV systems is the *open-circuit voltage* or V_{oc}. Open-circuit voltage, shown in Figure 3.17, is the voltage produced by a PV cell, module, or array with no load (inverter or battery) connected. That is, the leads are not connected to a load. As a result, no current can flow.

Placing the probes of a voltmeter across the leads of PV module, allows one to test the open-circuit voltage. Voltage, as you learned earlier, is a measure of electromotive force, the force that makes electrons move through a wire. Voltage can occur without current flowing. If you've ever measured a battery's voltage, all you are measuring is the potential difference between the positive and negative ends of the battery. No current is flowing through the voltmeter.

Open-circuit voltage, on the other hand, represents the maximum voltage a PV module could produce under standard test conditions. Before solar systems are installed, installers must calculate the open-circuit voltage of the array. That's done by simply multiplying the open-circuit voltage of the modules by the number of modules. Ten modules, each with an open-circuit voltage of 36 volts produce an array with V_{oc} of 360 volts. That's the highest voltage you will ever see from a PV array except on extremely cold, sunny days. (More on this shortly.)

Open-circuit voltage is important because it allows an installer to select certain equipment needed for a system, for example, the inverter or charge controller. That's because all inverters are designed to operate within a specific voltage range, usually about 150 to 550 volts for residential inverters. To prevent damage, the maximum input voltage rating of inverters and charge controllers must be higher than the maximum output voltage of the array. If it isn't, you'll smoke them—that is, damage them. Smoke will emanate from the equipment, and they'll stop working.

As just noted, the output voltage of a module or array can be higher on cold days. That's because the voltage of a PV device increases as temperature decreases. The maximum output voltage of the array is therefore calculated by multiplying the V_{oc} of the array by a temperature adjustment factor based on the coldest temperature for the site. (Most installers use a midpoint between the coldest temperature ever experienced and the average low temperature. That usually occurs in January in the Northern Hemisphere.)

Figure 3.17 illustrates how the next three parameters, maximum power voltage, maximum power current, and maximum power point are calculated. As illustrated, the *maximum power point* or P_{mp} is the point on the I-V curve at which the power output (measured in watts) reaches its highest point under standard test conditions. This is also referred to as the module's *rated power*. For a 300-watt module, the maximum power output under STC is 300 watts.

The output of a module in watts can be calculated by multiplying the amps it produces under a certain set of conditions by the voltage. I use the equation watts = amps × volts ($W = A \times V$). As shown in Figure 3.17, then, the maximum power current (rated power) is the product of the maximum power voltage (V_{mp}) and maximum power current (I_{mp}) at STC. Put another way, V_{mp} and I_{mp} are two points on the x and y axes that, when multiplied together, yield the maximum power, or P_{mp}.

The *maximum power voltage* and *maximum power current* are not particularly important when sizing PV systems. However, if you become an installer, you will use them to calculate efficiency (line loss or voltage drop that occurs as electricity travels through the conductors from your array to your inverter). That said, maximum power voltage and maximum power current are extremely important for getting the most power from a PV array. Why's that?

As you'll learn in Chapter 6, most inverters and charge controllers contain circuitry that ensures that the array operates at its maximum power point—no matter what the irradiance or temperature. This function is known as *maximum power point tracking* (MPPT). It is explained in the sidebar "PV Array Voltage, String Size, and Choosing the Correct Inverter" in Chapter 6. For now, all you need to know is that the inverter or charge controller in battery-based systems monitor voltage and amperage of the array and adjust them so the array operates at all times at the maximum power point. MPPT will increase the output of a PV array by about 15%

per year. The greatest benefits come when you may need it the most—during the winter when the PV cells are cold, the array voltage is high, and the availability of sunlight is lowest.

Another parameter that is listed on a module spec sheets and the labels affixed to the back of modules is the *maximum system voltage*. It is the highest voltage to which an array can be wired when using that module. In residential systems, the National Electric Code limits maximum system voltage to 600. The maximum system voltage rating of the modules must be 600 volts. In commercial systems, the limit is 1000 volts. This provision of the National Electric Code (NEC) allows installers to connect more modules in series to produce a higher-voltage system. The higher the voltage, the more efficient a system operates. (High-voltage electricity reduces line loss.)

Another parameter listed on module spec sheets and on modules is the *series fuse rating*, which is the value of the fuses or circuit breakers that must be installed in the DC wires that run from the array to an inverter. (The series fuse rating for most modules in residential arrays is 15 amps.)

Series fuses are typically installed in combiner boxes at the array or in the inverters (Figure 3.18). A combiner box is a device that allows an installer to combine all of the output of two or more series strings of modules into one set of wires. The series fuse protects the wire running from the array to the inverter or charge controller in a battery-based system. However, it also protects modules from excessive reverse current—that is, current over 15 amps that could flow from equipment like an inverter back through an array. This is a highly abnormal condition, but installers must protect modules from this remote possibility. The modules cannot handle more current.

While all these numbers may seem confusing at first, as you learn how to size a system, they'll eventually make more sense.

Durability and Fire Resistance

Module prices have fallen dramatically in recent years, in part, because of cost-cutting measures in manufacturing. But, as with so many products, lower cost may translate into lower quality. Most module failures result from breaks in the thin silver ribbons and solder used to connect cells in a module. As a result, several manu-

FIGURE 3.18. Series Fuses/Breaker in Combiner Box. This combiner box contains two fuses, one for each series string. These breakers protect the modules from reverse current and protect the wires as well, as explained in the text. In a combiner box, all the negative wires from the series strings are combined at a bus bar and all the positive wires are combined at the breakers. Combining the positive and negative wires is known as parallel wiring. Parallel wiring increases the amperage but does not affect the voltage. Credit: Dan Chiras.

facturers are developing third-party testing and higher standards of durability and fire protection.

Although it may take time for this process to improve, module manufacturers can subject their products to accelerated testing through a number of independent testing facilities, like Underwriters Laboratories.

When it comes to durability, one of the most common questions I'm asked about PV modules is: "How do modules hold up against hail?" The answer is: pretty well. In fact, PV modules are designed to withstand a 25 mm (just less than 1 inch) diameter hail stone striking them at 51 miles per hour (23 mps). Faced with increasingly more violent storms, manufacturers are developing even more demanding tests with much larger hail stones striking at higher rates of speed. How prevalent is hail damage?

In 2013, I inspected over 10,000 modules in 100 PV installations in western Missouri and found only a handful of modules damaged by hail—and these modules were mounted flat on a roof, making them more vulnerable. The impact craters looked like they were the result of strikes by softball-sized hail.

Modules are also tested for fire resistance. This rating system has recently been improved to take into account differences in module construction—for example, whether a module is encased in glass (glass front and back) or glass in the front and plastic in back. It also takes into account the thickness of the glass. Fire ratings also factor in the type of roof and the type of rack the modules are mounted on, a topic beyond the scope of this book.

Advancements in PV: What's on the Horizon?

Although PVs have come a long way since Becquerel's discovery in 1839, researchers continue to advance the technology. They are constantly looking for ways to improve the efficiency and reduce the cost of solar electric technologies to make it more affordable to customers and profitable for businesses. This effort has led to some rather promising developments in recent years. Some of these technologies are already available; others are not ready for prime time, but could be manufactured commercially in the near future. Researchers have also developed novel applications of solar electricity like building-integrated PV (BIPV, e.g., solar roof tiles) and ways to modify arrays to increase their output. Let's start with BIPVs.

Building-integrated PVs

BIPV incorporates photovoltaic material (and therefore, power-generating capacity) into the components of a building envelope, for example, the roof, windows,

FIGURE 3.19. (*a*) Solar Roof Tiles. Solar roof tiles like the ones shown here can be used in place of roof shingles on sloped roofs. They protect the roof from rain and snow much like an ordinary roofing shingle or tile but produce electricity as well. Credit: Sanyo.
(*b*) Solar Used for Standing Seam Metal Roof. This photo shows solar laminate, or PVL, from Uni-Solar made from amorphous silicon. This product can be applied on new or existing standing seam metal roofs. Credit: Energy Conversion Devices/Uni-Solar.

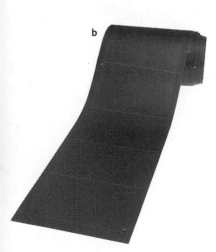

skylights, and exterior walls. Many BIPV products incorporate some form of thin-film solar material.

On commercial buildings with flat roofs, for example, you can install a flexible roofing membrane coated with thin-film silicon. For pitched roofs, you can install solar roof tiles—PV modules that replace conventional roof tiles. Solar roof shingles made from thin-film material were produced for a while by one manufacturer (Figure 3.19a). While solar roof shingles may seem like a neat idea, they came with some serious limitations. For one, to install the 7-foot (2-meter) shingles, numerous holes had to be drilled in roofs to wire one shingle to the next. When an installer was done, a roof may have had 300 to 400 holes in it—not a terribly great idea, as the main operating principle during roof installations is to minimize penetrations. In a couple instances, wires in the system shorted, causing house fires. As a result, the product was pulled from the market.

Experience with solar shingles reinforces my first rule of technology: Just because an idea is cool, that doesn't mean it is a smart idea. The company that manufactured roof shingles has subsequently gone belly up. To my knowledge, no one is producing solar roof shingles.

On standing seam metal roofs, you can install a flexible thin-film material known as a *PV laminate*, or *PVL*, like the one shown in Figure 3.19b. PV laminate is made by spraying thin-film material on a flexible backing such as a thin layer of aluminum or steel. It is then coated with UV-stabilized polymers (plastic). The product is manufactured in rolls. When it is time to apply PVL, a backing sheet is peeled off the laminate, revealing a sticky surface that adheres the PVL to standing seam metal roofing.

PVL can be used for parking structures and the metal roofs of barns, homes, schools, government buildings, and so on. It is best applied to brand new roofs—actually, it is best to put it on the metal roof panels before they are even installed on a roof.

As noted earlier, glass in skylights and windows can also be coated with a thin-film solar material to generate electricity. Bear in mind that glass is typically mounted vertically. North-facing glass will receive no direct sunlight for half the year. East- and west-facing glass will be in shade half of each day. Only south-facing glass will be sufficiently illuminated. If that glass is protected by overhangs in the summer, however, solar gain will be reduced. Solar gain in south-facing glass will also be reduced throughout the summer because of its vertical orientation. This minimizes solar gain, for the reasons discussed in Chapter 2.

Solar paints have received a significant amount of press in recent years. These are paints laced with photovoltaic material. Again, while a cool idea, they would suffer the same problems as solar glass.

Solar modules can themselves be mounted in specially designed panels in the exterior walls of buildings—even skyscrapers—although energy production will be limited for reasons of solar orientation just discussed.

As noted earlier in the chapter, some manufacturers sandwich crystalline silicon cells in glass panels that are used for the roofs of bus stop shelters or as canopies.

Conventional PV modules can be used to make awnings, windows, carports, parking structures, and backyard shade structures. To me, this is the most promising BIPV. If designed correctly, structures such as these can contribute a significant amount of electricity to a home or business.

BIPVs are being used primarily in new buildings, although they can be incorporated into existing homes and office buildings. They offer several advantages over more conventional PV technologies. One of the most important advantages is that they perform multiple functions. This reduces resource depletion and construction costs. For instance, solar awnings, like the ones shown in Figure 3.20, provide shade for windows, lower cooling costs, and eliminate the need for conventional awnings. They do all this while generating electricity.

Another advantage of BIPVs is that they tend to blend in better than conventional PV modules, which many people find appealing.

FIGURE 3.20. PV Awning. Solar awnings, like the one shown here, provide shade for windows and walls, eliminating the need for conventional awnings. Like other forms of building-integrated PVs, they perform more than one function. Awnings shade windows and walls and thus cool buildings in the summer and produce electricity year round. Credit: Lighthouse Solar, Boulder, CO.

Before you embrace a BIPV technology, however, think seriously about its orientation to the Sun throughout the year. If it can't be oriented properly, its output will suffer.

Bifacial Solar Cells

Yet another interesting technology is the *bifacial solar cell*, also known as a *hybrid solar cell*. Bifacial solar cells consist of a monocrystalline silicon wafer sandwiched between two ultrathin amorphous silicon layers. Because of this, these modules can capture sunlight falling on both sides of their cells. (That's the reason they're called *bifacial* modules.)

In a bifacial module, the front side of the panel generates electricity from direct and diffuse solar radiation (Figure 3.21). The backside generates electricity from diffuse light from the sky as well as reflected (diffuse) light bouncing off surrounding surfaces. Under the right conditions, bifacial modules can produce more power than conventional modules. Bifacials range in efficiency from 20.5% to 22.9%, although output is highly dependent on how they are mounted. If they're mounted on a dark-colored roof, for instance, don't expect a very significant increase in production. Tread carefully here.

FIGURE 3.21. (*left*) Bifacial modules like those installed at the Midwest Renewable Energy Association harvest solar energy off both sides of the panel, although the front side shown in this photograph is by far the most productive source of electricity. (*right*) Sunlight reflecting off the light-colored roofing illuminates the array, boosting its output. Credit: Dan Chiras.

When considering this or any other improved design, always be sure to compare the new type of solar module to conventional modules on the basis of the cost per kilowatt-hour of electricity. What you will often find is that the higher cost of higher-efficiency modules does not make economic sense. In other words, lower-efficiency options are often still cheaper when compared on the basis of cost per kilowatt-hour of electricity. It often makes more sense to add a few more conventional modules to boost output of a PV array.

Concentrating PV

The efficiency of solar modules can be increased by using lenses and mirrors to concentrate sunlight on PV cells (Figure 3.22). Concentrating sunlight greatly increases the solar input to the cells. The more sunlight that can be focused on a PV, the more electricity it generates. Concentrating collectors must be mounted on trackers—devices that follow the Sun across the sky—to operate most efficiently.

Concentrating solar collectors are not a new idea. Early attempts in the 1970s, however, proved disastrous, as concentrated sunlight literally burned out PV cells. To avoid this problem, PV cells in concentrating collectors must be cooled. Cooling is achieved by applying a heat sink to the back of the solar cells. This usually consists of a fin-shaped piece of aluminum that draws heat off the solar cells.

Concentrating solar systems when combined with expensive multi-junction cells can achieve efficiencies of around 45%. Because they operate at a higher efficiency, they require less semiconductor material than other systems, making multi-junction cells economical to use in commercial power production facilities.

FIGURE 3.22. Concentrating Solar Collector. Solar collectors like these large commercial arrays use lenses to concentrate sunlight on PV cells, dramatically boosting their output. Credit: Amonix.

Concentrating systems do have some limitations. Although they operate very efficiently on clear, sunny days, they function very poorly—or not at all—on cloudy days. That's because clouds scatter incoming solar radiation, increasing the amount of diffuse light. Diffuse light cannot be concentrated like direct light. (To see this for yourself, try concentrating light with a magnifying glass on a cloudy day.)

Third-generation Solar Cells

In later chapters, I'll discuss other advances in solar electricity such as changes in modules, charge controllers, and inverters. In this chapter, I'll discuss new thin-film solar cells that are part of a *third-generation* solar electric technology. I've chosen to discuss three of the most promising designs: dye-sensitized solar cells, organic solar cells, and quantum dot solar cells.

Dye-sensitized Solar Cells

Dye-sensitized solar cell (DSSC) is a type of thin-film PV. As shown in Figure 3.23, DSSCs contain particles of titanium dioxide (TiO_2) coated by a light-absorbing dye. Sunlight enters the cell through a transparent top layer, striking the dye molecules. Photons with sufficient energy to be absorbed by the dye excite electrons in the dye molecules. These electrons are transferred directly into the conduction band of the titanium dioxide molecules. The electrons then move to the clear surface layer, which doubles as the contact.

Electrons lost from the dye molecules are rapidly replenished by iodide ions (I^-) in the electrolyte surrounding the titanium oxide/dye particles. When this occurs, iodide ions are converted to the oxidized state (I_3^-). Oxidized ions, in turn, recoup their missing electron by diffusing (moving) to the bottom of the cell. Here they pick up electrons flowing back into the cell via a conductor through the external circuit—just like an ordinary PV cell.

Unlike previous cells described in this chapter, DSSCs are not solid-state devices. They are *photoelectrochemical* cells. Although they rely on the photoelectric effect to generate electrons (in the dye) to create current, they also rely on ions in the electrolyte to transfer electrons to the dye molecules.

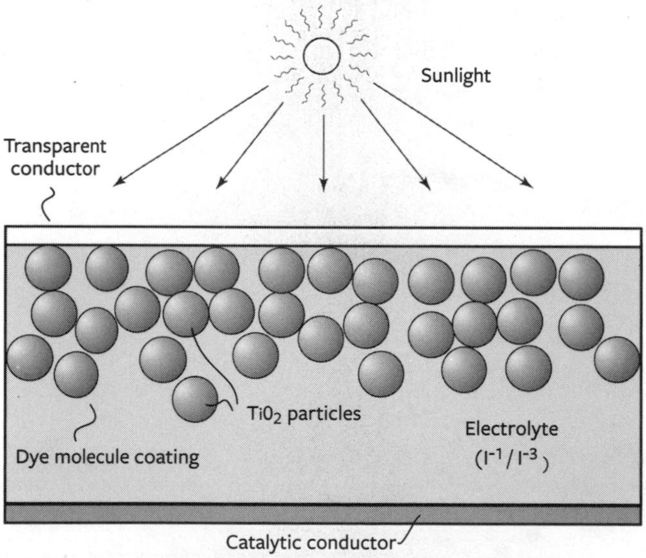

FIGURE 3.23. Dye-sensitized Solar Cell. This drawing shows the anatomy of one of the newest and most promising solar cells under development. Credit: Anil Rao.

Labels in figure: Sunlight; Transparent conductor; TiO$_2$ particles; Dye molecule coating; Electrolyte (I^{-1}/I^{-3}); Catalytic conductor

This technology uses low-cost materials and can be used to manufacture long semi-flexible and semi-transparent rolls. As a result, cells can be assembled using much less energy with much simpler and less expensive equipment than is required to manufacture conventional monocrystalline and polycrystalline PVs. Consequently, they should be cheaper to manufacture.

On the downside, manufacturers have had a tough time eliminating a number of expensive materials needed to manufacture this product (platinum and ruthenium). The liquid electrolyte is also subject to freezing, making it quite challenging to design a cell suitable for all climates. Another problem is low conversion efficiency. DSSCs are currently around 11% efficient, although efficiencies of 15% have been attained in the laboratory. Although the conversion efficiency is lower than the best thin-film PVs on the market, some believe that lower production costs could enable these cells to compete with electricity generated by coal and other fossil fuels within the next decade or so. After all, it is the cost per kilowatt-hour of electricity that determines how competitive it will be.

Since the previous edition of this book, dye-sensitized solar cells have made their way into the market as power sources attached to gym bags to charge portable electronic devices. Some believe that commercial production is just around the corner, but, as Yogi Berra said, "It's tough to make predictions, especially about the future."

Organic Solar Cells

Another technology that holds promise for the future is the *organic solar cell*. Organic solar cells consist of thin plastic films (typically 100–200 nm) containing organic (carbon-based) materials capable of emitting electrons when illuminated by sunlight to produce DC electricity like silicon-based cells. (Although they do it in a very different manner.) Some organic cells contain long chained polymers.

The organic compounds used in organic cells come with tongue-twisting names such as polyphenylene vinylene, copper phthalocyanine (which is a blue or green organic pigment), and carbon fullerenes. In the carbon fullerene cells, a polymer (plastic) releases electrons when struck by light. The neighboring carbon fullerene accepts the electrons.

The simplest organic cells consist of a single layer of photo-excitable and conductive organic material sandwiched between two conductors, as shown in Figure 3.24. Understanding how the cells generate electricity requires an advanced degree in physics or organic chemistry. If you're interested, you can find excellent descriptions on the internet.

Sunlight

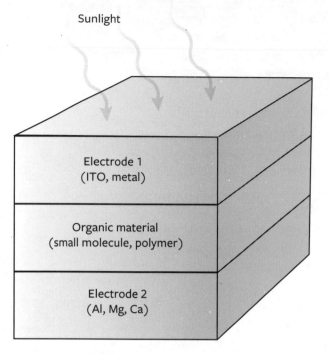

FIGURE 3.24. Single-layer Organic Solar Cell. The simplest (and least efficient) organic solar cell consists of a single layer of reactive material. Credit: Forrest Chiras.

The organic molecules used to make organic PVs can be produced in mass quantities and relatively inexpensively. The organic molecules currently being tested are also capable of absorbing a large amount of solar energy per unit of material, unlike other thin-film materials. In addition, slight chemical modifications of these molecules can result in a variety of compounds that absorb different wavelengths of solar radiation, which could aid in the production of very high-efficiency multi-junction cells. Because of this, most of the research is focused on creating cells that have multiple layers of different materials containing donor and acceptor molecules. Finally, because they use so little material and the materials are inexpensive, organic cells could eventually be produced at a very low cost. (That remains to be seen, however.)

Unfortunately, to date, most organic cells produced in the laboratory are rather inefficient. The highest efficiencies range from 6.5% to 11.6%. (Expect that to change.)

Organic solar cells have the additional drawback of breaking down easily in sunlight, although efforts are underway to address this issue. Some "experts" think that organic cells could eventually be produced at a cost as low as 10 cents per watt. Poly and multicrystalline solar cells cost about seven times more to produce.

Quantum Dot Solar Cells

Another interesting development is the *quantum dot solar cell*. This technology draws on the relatively new and exciting branch of science called *nanochemistry*. Like other third-generation solar cells, this technology relies on materials that absorb a variety of wavelengths of light. It's therefore ideal for multi-junction solar cells. Because they permit greater use of the solar spectrum, they are potentially higher-efficiency cells. So far, however, efficiencies in this new technology are fairly low—between 7% and 8.7%. (Again, expect those numbers to change.)

Quantum dots (QDs) are, as their name implies, tiny semiconducting particles. Their ability to absorb different wavelengths of light can be changed by altering their size. In the lab, the dots can be grown to a range of sizes, allowing them to absorb a variety of wavelengths. This occurs without any changes in the material.

Single-junction QD cells are made from lead sulfide (PbS). They can capture long-wave heat radiation to generate electricity, something not possible with traditional solar cells. (Remember that 50% of the Sun's output is infrared radiation.) This characteristic could dramatically increase the output of a solar cell. Some scientists believe quantum dot solar cells could eventually achieve efficiencies of nearly 70%.

Quantum dot solar cells should be easy and inexpensive to produce with relatively simple, inexpensive equipment. The dots are currently sputtered onto a substrate. In large-scale production facilities, they could be sprayed on like ink.

Conclusion: Should You Wait for the Latest, Greatest New Technology?

Knowing that new PV technologies are in the offing, many people ask us if they should wait a bit before they buy a solar electric system. Does it make sense to delay your installation until the new, more efficient PV technologies hit the market?

This is a fair question, but the answer is no.

Newer, more efficient solar cells are currently being used in modules. More efficient versions are also on their way. However, one of the key considerations when installing a PV system is not the efficiency of the PV modules, but the cost per kilowatt-hour of electricity they will produce over their lifetime.

Most installers will provide a cost based on the capacity of a system, that is, the cost per watt of installed capacity. At this writing (January 2016), grid-tied solar systems are typically running as low as $2.50 to as high $4.50 per watt in the United States. The national average is about $3.50 per watt. A 10,000-watt or 10-kW system would cost somewhere between $25,000 and $45,000. Batteries will typically add $2.00 per watt to the system cost.

This wide variation in price is due to a number of factors, such as local labor costs and the cost of materials, which can vary from one region to another.

While cost per watt of installed capacity is a helpful measure, the best way to determine the most cost-efficient option is to estimate the amount of electricity a system will produce over its lifetime, then divide the cost of the installed system by that projected output. Installers typically use 30 years as the lifetime of a PV system for cost calculations, but you will very likely get 30 to 50 years of service from a PV array (although you will probably have to replace your inverter once or twice during that period). For the longest service, be sure to invest in high-quality modules that are made to last. One manufacturer that's made tremendous strides in improving the longevity of their solar modules is SunPower. They've strengthened the connections and built-in "expansion capability" in the connections between individual cells to reduce breakage. They've also incorporated a thin copper mesh in the cells to accommodate contraction and expansion. They also embed the silver contacts inside the cell to reduce shading and prevent corrosion.

Efficiency matters if space is limited. If meeting your electrical needs requires a 10-kW system, and the only place you have with good solar access is your roof, you may need to install the more efficient—and expensive—modules.

While conversion efficiency may not matter to you, efficiency is driving the market, and over the long haul it will result in more efficient and hopefully less expensive modules. Those companies that can produce higher-efficiency modules at a lower cost with less raw material in an environmentally friendly manner stand to transform the way the world makes energy—and make billions in the process.

CHAPTER 4

Solar Site Assessment

Before you invest your hard-earned money in a solar electric system, it is important to determine whether a solar electric system makes sense. Will a PV system meet your needs? How much will it cost? Is the investment worth it? Would you be better off continuing to buy utility power? What nonmonetary benefits might you receive? What about alternatives to buying a PV system, in particular, community solar or leasing a system?

In this chapter, we begin by focusing on whether a PV system makes sense from an economic perspective. The decision to invest in a PV system requires an analysis of two factors: how much electricity you need and how much it will cost to meet your needs with solar electricity.

Assessing Electrical Demand

One of the most common questions I am asked is "How much is it going to cost to install a solar electric system on my home?"

The answer is, "That depends."

The cost of a solar electric system depends on many factors. One of the most important is that amount of electricity you consume. That will determine the size of the system and its cost. Other factors that come into play include the type of system, the distance the installer must travel, and the difficulty of the installation (e.g., how steep is your roof). Another factor that affects the cost are financial incentives. Are there any local, state, or federal tax credits?

Solar resources and electrical consumption vary, but most residential solar electric systems fall within the range of 8 to 15 kW. Energy misers can get by with less. For example, my home in Colorado is extremely energy efficient, consuming only a fraction of the electricity of a standard home. I was able to power it with a 1.2-kW PV array. Ed Begley, Jr., an actor and environmental advocate, powers his small energy-efficient California home as well as an electric car with a 6-kW system. Most residences, however, will require 8- to 15-kW systems. In an all-electric home equipped with an assortment of electric appliances—especially heavy hitters like central air conditioners, electric space heating, electric water heater, and electric stoves—average monthly electrical consumption typically falls within the 2,000 to 3,000 kilowatt-hours/month range. I've encountered people living in McMansions who consume over 5,000 kWh per month. Even in sunny climates, very large PV systems are required to meet their electrical demands. If you fit in this category, get ready to write a huge check. You might need a 15- to 30-kW system, maybe even larger.

How do you determine electrical energy consumption in a home? Well, how you go about calculating energy consumption depends on whether it's an existing structure or one that's about to be built. Let's begin with existing structures.

Average Annual Electrical Use

Based on national averages, US homes consume about 1,000 watts continuously, or about 24 kWh per day. That's 730 kWh per month or 8,760 kWh per year, which is about $876 worth of electricity a year (not counting meter-reading fees, taxes, and surcharges) at 10 cents per kilowatt-hour.

Assessing Electrical Demand in Existing Structures

Assessing the electrical consumption of an existing home or business is fairly easy. Most people obtain this information from their monthly electric bills, going back two to three years, if possible. (In many areas, every utility bill includes a summary of year-to-date electrical consumption.) If you don't save your electric bills, a telephone call to the local power company will usually yield the information you need. Some companies will allow you to access the data online through their websites. All you need is your customer number.

If you've purchased a home that's been around for a while, you can obtain energy data from the previous owner—if they've saved their utility bills and are willing to share this information with you. Or, they may be willing to contact the local utility company on your behalf to request a summary of their electrical consumption over the past two to three years. Remember, however, a house does not consume electricity, its occupants do, and we all use energy differently. If a previous homeowner and his or her family used energy wastefully, their energy consumption data may be of little value to you if you are an energy miser.

To determine total annual electrical consumption from utility bills, don't look at the cost in dollars and cents, look instead for the kilowatt-hours of electricity consumed each month. If you're installing a grid-tied solar electric system, annual electrical demand is all you need. You can size your system based on this information. (I'll show you how, shortly.)

If you are thinking about severing your ties with the utility—that is, taking your house off the grid—you will need to determine monthly averages—that is, how much electricity is used, on average, during each month of the year. For example, if your records go back four years, add the electrical consumption for all four Januarys, and then divide by four. Do the same for February and each of the remaining months. Record monthly averages on a table. This will permit you to determine when electrical demand is the greatest. Off-grid systems are sized to meet demands during the times of highest consumption.

After you have calculated annual energy consumption, in the case of a grid-connected PV system, or monthly averages, in the case of an off-grid system, take a few moments to look for trends in energy use. Is energy consumption on the rise or is it staying constant or declining? If you notice an increase in electrical use in recent years, for instance, because you've installed a big-screen TV or added air conditioning, dump earlier energy data and recalculate the averages based on the most recent bills. Early data will artificially lower the average, and the averages won't represent current consumption. You could end up undersizing your PV system.

If, on the other hand, electrical energy consumption has declined because you've replaced your inefficient furnace and refrigerator with newer, more frugal models, earlier data will artificially inflate electrical demand. You'll need less electricity than the averages indicate—and a smaller system.

Assessing Electrical Demand in New Buildings

Determining monthly electrical consumption is fairly straightforward in existing homes and businesses. Determining electrical demand in a new home—either one that's just been built or one that is about to be built—is much more difficult.

One method used to estimate electrical consumption is to base it on the electrical consumption of your existing home or business. For instance, suppose you are building a brand new home that's the same size as your current home. If the new home will have the same amenities and the same number of occupants, electrical consumption could be similar to your existing home.

If, however, you are building a more energy-efficient home and are installing much more energy-efficient lighting and appliances and incorporating passive solar heating and cooling (all of which I highly recommend), electrical consumption could easily be 50%, perhaps 75%, lower than in your current home. If that's the case, adjust electrical demand to reflect your new, more efficient lifestyle.

Another way to estimate electrical consumption is to perform a load analysis. A load analysis is an estimate of electric consumption based on the number of electronic devices in a home, their average daily use, and energy consumption. It is a lot more difficult to calculate than you would expect.

To perform a load analysis, a homeowner begins by listing all the appliances, lights, and electronic devices that will be used in his or her new home. Rather than list every light bulb separately, however, you may want to lump them together by room. I like to work with clients one room at a time, using a spreadsheet I've prepared for each project. Or, you can use a worksheet like the one shown in Table 4.1. Similar worksheets can be found online at Northern Arizona Wind and Sun's website (solar-electric.com) and other websites.

Once you've prepared a complete list of all the devices that consume electricity, your next assignment is to determine how much electricity each one uses. The amount of electricity consumed by an appliance, lamp, or electronic device can be determined by consulting a chart like the one in Table 4.2. Charts such as this list

Table 4.1. Load Analysis

Individual Loads	Quantity	× Volts	× Amps	=	Watts AC	Watts DC	× Use (hrs/day)	× Use (days/wk)	÷ 7 days	=	Watts Hours AC	Watts Hours DC
									7			
									7			
									7			
									7			
									7			
									7			
									7			
									7			
									7			
									7			

AC Total Connected Watts: _____   **AC Average Daily Load:** _____

DC Total Connected Watts: _____   **DC Average Daily Load:** _____

typical wattages for a wide range of electrical devices. Detailed listings are available online. Better yet, log on to WE Energies's website. It contains a nifty energy calculator that allows you to estimate electric consumption room by room (find it at webapps.we-energies.com/appliancecalc/appl_calc.cfm).

Table 4.2. Average Electrical Consumption of Common Appliances

General household		Kitchen appliances		Entertainment	
Air conditioner (1 ton)	1500	Blender	350	CB radio	10
Alarm/Security system	3	Can opener (electric)	100	CD player	35
Blow dryer	1000	Coffee grinder	100	Cellular telephone	24
Ceiling fan	10–50	Coffee pot (electric)	1200	Computer printer	100
Central vacuum	750	Dishwasher	1500	Computer (desktop)	80–150
Clock radio	5	Exhaust fans (3)	144	Computer (laptop)	20–50
Clothes washer	1450	Food dehydrator	600	Electric player piano	30
Dryer (gas)	300	Food processor	400	Radio telephone	10
Electric blanket	200	Microwave (.5 ft³)	750	Satellite system (12 ft dish)	45
Electric clock	4	Microwave (.8 to 1.5 ft³)	1400	Stereo (avg. volume)	15
Furnace fan	500	Mixer	120	TV (31.5-inch color)	22–30
Garage door opener	350	Popcorn popper	250	TV (40-inch color)	28–37
Heater (portable)	1500	Range (large burner)	2100	TV (48-inch color)	35–55
Iron (electric)	1500	Range (small burner)	1250	VCR	40
Radio/phone transmit	40–150	Trash compactor	1500		
Sewing machine	100	Waffle iron	1200	**Tools**	
Table fan	10–25	**Lighting**		Band saw (14″)	1100
Waterpik	100	Incandescent (100 watt)	100	Chain saw (12″)	1100
		Incandescent light (60 watt)	60	Circular saw (7¼″)	900
Refrigeration		Compact fluorescent	16	Disc sander (9″)	1200
Refrigerator/freezer	540	(60 watt equivalent)		Drill (¼″)	250
22 ft³ (14 hrs/day)		Incandescent (40 watt)	40	Drill (½″)	750
Refrigerator/freezer	475	Compact fluorescent	11	Drill (1″)	1000
16 ft³ (13 hrs/day)		(40 watt equivalent)		Electric mower	1500
Sun Frost refrigerator	112			Hedge trimmer	450
16 ft³ (7 hrs/day)		**Water Pumping**		Weed eater	500
Vestfrost refrigerator/freezer	60	AC Jet pump (¼ hp)	500		
10.5 ft³		165 gal per day, 20 ft. well			
Standard freezer	440	DC pump for house	60		
14 ft³ (15 hrs/day)		pressure system (1–2 hrs/day)			
Sun Frost freezer	112	DC submersible pump	50		
19 ft³ (10 hrs/day)		(6 hours/day)			

For more accurate data, I strongly recommend that you check out the nameplates on the appliances and electronic devices you'll have in your new home. The nameplate is a sticker or metal plate that lists the unit's electrical consumption. The measure you are looking for is watts. As you may recall, wattage is a measure of instantaneous power consumption. Manufacturers sometimes list amps and volts. If the nameplate lists amps, but not watts, calculate wattage by simply multiplying the amps × volts (amps × volts = watts).

Multiplying amps by volts yields the wattage of many electronic devices, among them resistive devices, such as electric heaters and electric stoves. (A resistive device is one in which electricity flows through a metal that resists the flow of electrons. This, in turn, produces light and heat.) It also works for universal motors. Universal motors are typically found in smaller devices such as vacuum cleaners, blenders, and small electric tools. For devices with induction motors, such as fans, washing machines, clothes dryers, dishwashers, pumps, and furnace blowers, multiplying amps by volts significantly overestimates the wattage. For these devices, multiply watts (the product of amps and volts) by 0.6 to obtain a better estimate of true wattage. (For a description of the rationale behind this derating, see the accompanying sidebar, "Real Power, Apparent Power, and Power Factor.")

Real Power, Apparent Power, and Power Factor

To calculate power consumption (watts) of electrical resistance devices, such as electric stoves or computers, simply multiply the voltage by the amps (power [watts] = volts × amps). However, for devices including certain electric motors (known as induction motors), transformers, and fluorescent lamp ballasts, the product of volts and amps is actually a bit higher than the actual wattage. In such instances, watts × volts is known as the *apparent power*. The discrepancy is due to the fact that inductors and capacitors store energy as *current*, and voltage waveforms *alternate* from positive to negative. This stored energy causes the current waveform to be out of step with the voltage waveform. When the current waveform is out of step with the voltage waveform, more current is required to deliver the same amount of power. The factor that accounts for this difference between apparent power and real power is the *power factor*.

Power factor is the real power divided by the apparent power. Power factor = watts/(volts × amps). Power factor is always a number between zero and one, but can also be expressed as a percent. When calculating the real power (watts) for a device that has inductors or capacitors, you must multiply voltage times amps times power factor. Power = volts × amps × power factor. Unfortunately, you won't find power factor listed on appliance nameplates, but a reasonable estimate for most household devices containing inductors or capacitors is 0.6 or 60%.

Some experts reduce the wattage of other devices, for example, televisions, stereos, and power tools, below their listed power consumption as well. They do this because the listed wattage is not representative of the typical run wattage—that is, the wattage an appliance will draw when in normal operation. Put another way, the wattage listed on an appliance nameplate is the power the device draws at maximum load. Most devices rarely operate at maximum load, so the nameplate power overstates the actual power used. Reducing the nameplate wattage by 25% is a reasonable adjustment for such devices, although typical operating wattage may be even lower. Be sure, when determining the run wattage of laptop computers that plug into a charger or transformer to use the wattage listed on the charger, not the device. The same applies to cordless drills, cordless phones, and electronic keyboards.

A more accurate way to determine the wattage (power) of an electronic device is to measure it directly using a meter like those shown in Figure 4.1. To use it, plug the meter into an electrical outlet and then plug the appliance into the outlet on the face of the device. A digital readout indicates the instantaneous power (watts). When measuring a device that cycles on and off, such as a refrigerator, or one that

Plug-in Watt-hour Meters

The Kill A Watt and Watts Up? meters make it easy to measure power (watts) and energy (watt-hours) used by plug loads: lights, refrigerators, freezers, computers, TVs, etc. Just plug the meter into a regular 120-volt outlet and plug the appliance into the meter. Both meters read power (watts), volts, amps, and elapsed time (used for calculating average power use over time). These meters, however, are not designed for measuring electrical consumption by 240-volt appliances.

Power meters are useful for estimating load so you can size a PV system but they are also useful for gaining an appreciation for the energy demands of household appliances. Knowing which devices consume the most energy in a home may help you devise a strategy to reduce energy consumption. Surveying the energy use of household appliances could be a great educational project for children.

Both meters are available in a range of models, some with advanced features. The Kill A Watt meter's list price is about $40; the Watts Up? meter's list price is about $135. Both are available online at a discount. As an alternative to purchasing a meter, some public libraries and utilities have plug-in watt-hour meters available for loan.

FIGURE 4.1. (*a*) The Kill A Watt and (*b*) Watts Up? meters can be used to measure wattage of household appliances and electronic devices. Of the two, the Watts Up? is the more sensitive and allows for measurement of tiny phantom loads as well. Credit: (*a*) Dan Chiras, (*b*) Electronic Educational Devices.

has a varying load, leave the watt-hour meter connected for a week or two. The meter will record the total energy used during this period in watt-hours and will tell you how many hours you have been recording data. You can then calculate the number of kilowatt-hours the device consumes in a 24-hour day.

Power consumption can also be determined by checking spec sheets for various electronic devices. You can even go online to find them. Most will list key electrical parameters.

After you have determined the wattage of each electrical device, you must estimate the number of hours each one is used on an average day and how many days each device is used during a typical week. From this information, you calculate the weekly energy consumption of all devices in your home or business. You then divide this number by seven to determine your average daily consumption in watt-hours. You will use this number to determine the size of your PV system.

Load analysis may seem simple at first glance, but it is fraught with problems. One shortcoming I've encountered is that it is often difficult for clients to estimate how long each appliance runs on a typical day. Most people have trouble estimating how many hours or minutes they operate their toasters or blenders each day. Is it three minutes, five minutes, or ten? While some may be able to provide a fairly accurate estimate of how long the kitchen light is on, they haven't the foggiest idea how long the refrigerator runs or how many minutes they run the microwave. Most people tend to underestimate TV time, too. And what about the kids? Do they leave the lights on when you're not around? How many hours of television do they watch each day? How many hours a day do they spend on their computers?

Another problem with this process is that run times vary by season. Electrical lights, for example, are used much more in the winter, when days are shorter, than in the summer. Furnace blowers operate a lot during the winter, but not at all or very infrequently the rest of the year. So what's the average daily run time for the furnace blower?

Another problem with this approach is that many electronic devices draw power when they're off. Such devices are known as a *phantom loads*. They include television sets, VCRs, satellite phone receivers, cell phone and laptop computer chargers, and a host of other common household devices. Any electronic device that has an LED light that shines all the time or comes with a remote and instant-on feature will have a phantom load. At times, phantom loads can be quite significant. In years past, I have found that satellite receivers draw nearly as much power when they're off as when they're on (Figure 4.2). My new DirecTV satellite receiver uses 24 watts when it is off, and 26 when it is on. Phantom loads typically account for 5% to 10% of the monthly electrical consumption in US homes. If they're not factored into the load analysis, estimates can be off.

Phantom loads may be quite small, but they add up over time. Plug-in watt-hour meters can be used to measure phantom loads—if you let them run long enough for the kWh reading to register on the display.

Because of these problems, homeowners often grossly underestimate their electrical consumption. So, if you estimate power consumption, be sure to include phantom loads and be overly generous with your estimates.

As a side note, you may have noticed that the worksheet shown in Table 4.1 has a column labeled DC. This is for those folks who want to install an off-grid system and power some or all of their loads with DC electricity. DC wattages are also the product of amps × volts. A device that uses 1 amp at 48 volts requires 48 watts.

Once you've calculated daily electrical energy use, it's time to size a solar electric system, right?

Actually, no. Before you size a system, it's a good idea to look for ways to reduce electrical use through energy efficiency and conservation measures. That's because the lower the energy demand, the smaller the solar electric system you'll need. The smaller the system, the less you'll spend on your system. Smaller systems also require fewer resources to produce, so you're helping reduce your environmental impact, including your carbon footprint.

Money spent on energy-efficiency measures may also have a better return on investment than money spent on a PV system. An energy-efficient refrigerator, for instance, that saves you 1,000 kWh a year might cost $800 to $900. A PV array that would generate 1,000 kWh a year will cost you $2,500 to $4,500, depending on where you live. Even with the 30% federal tax credit, which now applies to systems installed before December 31, 2019, energy efficiency still trumps solar electricity when it comes to cost and sustainability.

FIGURE 4.2. Phantom Load. This superefficient, 50-inch Samsung flat-screen television uses only 60 to 130 watts when operating. Because of the instant-on feature, however, it continues to draw 8 watts when turned off. The satellite receiver attached to the TV uses 26 watts when operating and 24 watts when turned off. To save energy, homeowners can plug electronics like this into power strips that are shut off when the unit is not on. Or, a homeowner can install an outlet controlled by a switch; however, these must be installed during construction. Credit: Dan Chiras.

Because efficiency is both economically and environmentally superior to increasing the capacity of a renewable energy system, a few really conscientious installers recommend energy conservation and efficiency measures first. If you agree to pursue these measures, they can downsize your system, which could save you thousands of dollars. How do you determine the most cost-effective measures to reduce electrical consumption?

One way is to hire a professional solar site assessor. For a relatively small fee, usually $300 to $600, depending on travel time and complexity of the analysis, a solar site assessor will provide a list of ways you can trim your energy demand, evaluate the potential of your site for a solar electric system, tell you what kind of system will best meet your needs, recommend placement of the array and the rest of the equipment, and size a system. He or she will also provide an approximate cost and perform an economic analysis. (To locate a solar PV site assessor in your area, visit the Midwest Renewable Energy Association's website.)

Another option is to hire a home energy auditor. A qualified home energy auditor will perform a more thorough energy analysis of your home and make recommendations on ways you can reduce your demand. They will also prioritize energy-saving measures. They won't be able to give you any advice on solar, however. Alternatively, you can perform your own home energy audit. For guidance, check out my book, *Green Home Improvement.*

Most homeowners contact solar installers to evaluate their site and make recommendations. In my experience, very few installers will offer advice on energy conservation or energy efficiency. They are there to sell you a solar system—and the bigger the better. Some inexperienced or unscrupulous installers will also recommend installing PV systems in less-than-optimum sites—for example, on east- or west-facing roofs (when south-facing roofs are small, nonexistent, or are shaded). That's why an independent solar site assessor is such a good investment. They are working for you and not lining their pockets with the profit they make on an underperforming solar array.

Conservation and Efficiency First!

Before you invest in a solar electric system, the first step should be to make your home—and your family—as energy efficient as possible. Even if you're an energy miser, you may be able to make significant cost-saving cuts in energy use.

Waste can be slashed many ways. Interestingly, though, the ideas that first come to mind for most homeowners tend to be the most costly: new windows and energy-efficient washing machines, dishwashers, furnaces, or air conditioners (Figure 4.3).

While vital to creating a more energy-efficient way of life or business, they're the highest fruit on the energy-efficiency tree—and the most expensive.

Before you spend a ton of money on new appliances or better windows, I strongly recommend that you start with the lowest-hanging fruit. These are the simplest and cheapest improvements and yield the greatest energy savings at the lowest cost.

Huge savings can be achieved by changes in behavior. You've heard the list a million times: turning lights, stereos, computers, and TVs off when not in use. Turning the thermostat down a few degrees in the winter and wearing sweaters and warm socks. Turning the thermostat up in the summer and running ceiling fans. Opening windows to cool a home naturally, especially at night, and then closing the windows in the morning to keep heat out during the day. Drawing the shades or blinds on the east, south, and then west side of your house as the summer sun moves across the sky can help reduce cooling costs. All these changes cost nothing, except a little of your time, but can reap enormous savings—not just in your monthly energy bill, but also in the cost of a PV system.

FIGURE 4.3. Energy-efficient Washing Machine. Spending a little extra for an energy-efficient front-loading washing machine like this will reduce the size of your solar system. To dry clothes, though, consider using a solar clothes dryer (commonly called a *clothesline*). Credit: Frigidaire.

Other low-hanging, high-yield fruit includes boosting insulation levels and weather-stripping and caulking to seal leaks in the building envelope—tightening up and bundling up our homes and workplaces. These measures reduce heat loss in the winter and heat gain in the summer. They not only reduce fuel bills, they dramatically increase comfort levels. Unfortunately, they are very rarely viewed as a high priority.

Because our homes are so leaky, sealing up the leaks can make a huge difference. If you add up all the tiny leaks in a building envelope around doors and windows and where plumbing and electrical penetrate a home, they'd be equivalent to a 3-foot by 3-foot (0.9 × 0.9 meter) window open 24 hours a day, 365 days a year. Yes, that's right. No one in their right mind would leave a window open all year long, but that is, in essence, what we're doing by not sealing up all those energy leaks in our homes.

FIGURE 4.4. Energy Star Label. When shopping for electronic devices such as computers, stereo system components, and television sets, be sure to look for the Energy Star label. This label indicates that the product is one of the most energy efficient in its category.

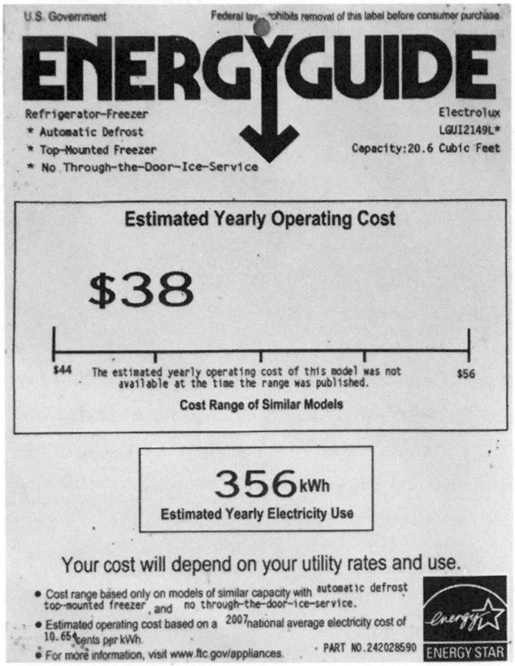

ENERGYGUIDE

U.S. Government

Federal law prohibits removal of this label before consumer purchase.

Refrigerator-Freezer
* Automatic Defrost
* Top-Mounted Freezer
* No Through-the-Door-Ice-Service

Electrolux
LGUI2149L*
Capacity:20.6 Cubic Feet

Estimated Yearly Operating Cost

$38

$44   The estimated yearly operating cost of this model was not $56
available at the time the range was published.

Cost Range of Similar Models

356 kWh
Estimated Yearly Electricity Use

Your cost will depend on your utility rates and use.

• Cost range based only on models of similar capacity with automatic defrost
  top-mounted freezer , and no through-the-door-ice-service.
• Estimated operating cost based on a 2007 national average electricity cost of
  10.65 cents per kWh.
• For more information, visit www.ftc.gov/appliances.   PART NO.242028590

ENERGY STAR

FIGURE 4.5. Energy Guide Tag. When shopping for appliances such as refrigerators and freezers, be sure to look for the Energy Guide, like the one shown here. These guides indicate how much electricity an appliance will use in one year and how the appliance you are looking at compares to others in its product category. This one is for the refrigerator/freezer that we purchased for our home; it uses the least energy of all the Energy Star-rated refrigerators in its category. Credit: Dan Chiras.

Once you've tackled these simple, relatively inexpensive steps, it's time to consider more costly, slightly bigger-ticket items. The next step is to boost the insulation in your home. Insulate windows with insulated shades. Insulate attics, floors above crawl spaces or garages, and walls. An energy consultant can help you figure out the best ways to seal and insulate your home.

Another big-ticket item that can help is replacing your refrigerator. In many homes, refrigerators are responsible for a staggering 25% of the total electrical consumption. If your refrigerator is old and in need of replacement, unplug the energy hog, recycle it, and replace it with a more energy-efficient model.

Thanks to dramatic improvements in design, refrigerators on the market today use significantly less energy than refrigerators manufactured 15–20 years ago. An efficient 21 cubic foot fridge today will consume around 365 kWh per year. A 20-year-old fridge could easily consume four times that amount. Whatever you do, though, don't lug your old fridge out to the garage or take it down to the basement and use it to store an occasional case of soda or beer. It will rob you blind! The same goes for old freezers.

Waste can also be reduced by replacing energy-inefficient electronics with newer, considerably more efficient Energy Star televisions, computers, and stereo equipment (Figure 4.4). You can also trim some of the fat from your energy diet by installing more energy-efficient lighting, such as compact fluorescent lights or longer-lasting and even more efficient LED lights. (For more on Energy Star appliances see the accompanying sidebar, "Energy Star—and Beyond.")

Although efficiency has been the mantra of energy advocates for many years, don't discount its importance just because the advice has grown a bit threadbare. As it turns out, very few people have heeded the persistent calls for energy efficiency. And, many who have made changes have not fully tapped the potential savings.

Energy Star—and Beyond

Although all new appliances and electronics must comply with federal energy-efficiency regulations, those that exceed these standards display an Energy Star label. This label indicates that the appliance or electronic device is one of the most efficient in its product category. So, when it is time to replace a television or stereo—even a cordless phone or laptop computer—always look for models with the Energy Star label. As Table 4.3 shows, Energy Star refrigerators use 20% less energy than is federally mandated. Dishwashers use 41% less and clothes washers use 37% less energy than federal law requires.

Although buying an Energy Star-rated appliance or electronic device is a good idea, you can do better, sometimes much better. How?

Go to the Energy Star website and click on the appliance or electronic device listing. Here you'll find a list of all Energy Star-qualified models. The list indicates the percentage by which each device exceeds the minimum efficiency for that type of appliance.

As Table 4.3 shows, the best appliances on the list are considerably more energy-efficient than other Energy Star-qualified products.

If you shop from the Energy Star list, you will do well, but if you search the list for the most efficient models, you can do much better.

For an up-to-date list of energy-efficient appliances, US readers can log onto the EPA's and DOE's Energy Star website at www.energystar.gov. Click on appliances. Or, Consumer Reports has an excellent website that lists energy-efficient appliances as well. Their site also rates appliances on reliability, another key factor. Canadian readers can log on to oee.nrcan.gc.ca for a list of international Energy Star appliances.

You can also compare appliances by checking out the yellow Energy Guide posted on appliances. It will tell you how much electricity a particular appliance will use and how the model you are looking at compares to other models in that category. Figure 4.5 shows and example of an Energy Guide. In Canada, the same yellow tags are also posted on appliances, but it's called an EnerGuide. How cute is that?

When shopping for TVs and other electronics look for the Energy Star logo, but turn the darn things around and read the labels on the back. Manufacturers typically post maximum and typical wattages. You would be amazed at the variation. I found a 50-inch (127-cm) Toshiba LED flat screen that uses 61 watts. That's only 12 watts more than my efficient laptop.

Table 4.3. Energy Star Performance
(% better than federal mandated efficiency)

Appliance Type	Energy Star Criteria	Best on the List
Refrigerators	20%	53%
Dishwashers	41%	147%
Clothes Washers	37%	121%

Readers interested in learning more about making their homes energy efficient may want to read the chapters on energy conservation in one of my books: *Green Home Improvement* or *The Homeowner's Guide to Renewable Energy*.

Sizing a Solar Electric System

Once you've estimated your electrical demand and devised and implemented a strategy to use energy more efficiently, it is time to determine the size of the system you'll need to meet your needs. This is a step usually performed by professional solar installers. The cost of systems vary, of course, depending on the type of system. Although I'll elaborate on options in the next chapter, for now, it's important to note that PV systems fall into three categories: (1) grid-connected, (2) grid-connected with battery backup, and (3) off-grid.

A grid-connected system can be sized to meet some or all of your electrical needs. Excess electricity, if any, is fed back onto the grid, running the electric meter backward. The electric grid, in essence, serves as your battery bank. When you want that electricity back, it's yours for free. (Provided you use it within the parameters of your agreement with the utility. Some companies allow a homeowner or business to bank electricity for one month at a time. Others allow annual banking. More on this later.)

Grid-tied systems are the most popular. Nearly all systems installed these days fall into this category. There is a downside to them, as you will see in Chapter 5. That is, when the utility grid goes down, most grid-tied solar system also shut down, even on sunny days. (This feature prevents electricity from being fed onto a malfunctioning grid, which might be hazardous to utility company employees.) As you will learn in Chapter 6, the inverter manufacturer SMA has designed an inverter that will provide limited power during a power outage, but only when the Sun shines.

A grid-connected system with battery backup is similar to a grid-connected system, but has a battery bank to store electricity. These systems will continue to supply electricity if the grid goes down. But because the battery banks are typically small, these systems usually only supply electricity to critical loads like furnace blowers, well pumps, sump pumps, refrigerators, and the like—loads that are critical to you and your family.

As their name implies, off-grid PV systems are not connected to the utility grid. They're completely autonomous electric-generating systems. Surpluses are stored in batteries for use at night or on cloudy days.

Sizing a Grid-connected System

Sizing a grid-connected system is the easiest of all. To meet 100% of your needs, simply divide your average daily electrical demand (in kilowatt-hours) by the average peak sun hours per day for your area. As noted in Chapter 2, peak sun hours can be determined from solar maps and from various websites and tables (also, see the sidebar in this chapter, "Determining Peak Sun Hours").

As an example, let's suppose that you and your family consume 6,000 kWh of electricity per year (or 500 kWh per month). This is about 17 kWh per day. Let's suppose you live in Lexington, Kentucky, where the average peak sun hours per day is 4.5. Dividing 17 by 4.5 gives you the array size (capacity), 3.8 kilowatts. But don't run out and order a system quite yet.

This calculation yields system size if the system were 100% efficient and unshaded throughout the year—and was operating under standard test conditions. Unfortunately, no PV system is 100% efficient, and systems rarely operate under STC. As a result, most solar installers increase the size of grid-connected PV systems by around 22%. This accounts for high temperatures, which reduce system output; dust on modules; losses due to voltage drop as electricity flows through wires; resistance at fuses, breakers, and connections; inefficiencies of various components such as the inverter; and a few other factors. So, if your PV array is not shaded by trees or nearby buildings, you'd need a 20% to 25% larger system to provide 17 kWh of electricity per day. I use 22% in adjusting my calculations for all my grid-tied systems.

To calculate the array size using the 22% adjustment factor, you simply divide 3.8 kW by 0.78. The result is 4.87 kW. We typically round it up to 5.0 kW to be on the safe side, *provided the solar array is not shaded*. A 5-kW array would require eighteen 285-watt modules, nine in each series string.

Shading dramatically lowers the output of a PV system. Even a small amount of shade on a module can reduce its output. In olden days, the effect was quite dramatic. Newer modules have been designed with bypass diodes that permit electricity to flow around shaded areas. (It's a bit more complicated than that, but that's the idea.)

Budget-based Sizing

Solar installers will size a PV system based on budget—how much money a person has to spend—because not everyone can afford a PV system to meet 100% of their needs. Let's suppose a homeowner had $15,000 to invest in a PV system. To determine what he or she could get for $15,000, an installer will first determine the array size needed to meet the customer's needs, then calculate its cost at the prevailing rate. Let's use $3.50 per watt. In the previous example, a 5.3-kW array would cost $18,500. If a customer applied the 30% federal tax credit, the system would cost $12,985. Clearly, there would be enough money to purchase an array that would meet this customer's needs. If he or she only had $6,000, the customer would be able to purchase a system that met 46% of his or her electrical demand.

To determine the amount of shading on a PV system, professional solar site assessors and solar installers often use a Sun path analysis tool like the Solar Pathfinder shown in Figure 4.6. The Solar Pathfinder and similar devices determine the percentage of solar radiation blocked by local features in the landscape such as trees, hills, and buildings. These devices are used by solar installers to locate the sunniest (most shade-free) site on a property and, then, to determine how much shading, if any, will occur at that location throughout the year.

The Solar Pathfinder consists of a clear plastic dome over a grid that indicates the path of the Sun throughout the year at any given location. The device is set on a tripod at the proposed location of the PV array, pointed true south, and leveled. Once set up, reflections of trees and other objects that could shade an array appear on the plastic dome.

To estimate shading, the operator takes a photo of the reflection showing on the dome. The digital image is then downloaded into accompanying software, Solar Pathfinder Assistant. The installer traces the outline of the shade using the mouse. The program then determines the amount of shading and calculates the percent of sunlight that is available to the array each month of the year. It then estimates the output of the array based on shading.

FIGURE 4.6. Solar Pathfinder. This device allows solar site assessors and installers to assess shading at potential sites for PV systems, helping you select the best possible site to install an array. Credit: Shawn Schreiner.

FIGURE 4.7. Solar Pathfinder Dome Showing Shading. The dome of the Solar Pathfinder (not shown here) reflects all obstructions that will shade a solar array. A photograph of the dome is then entered into a computer program that allows the operator to trace the shading. The computer then calculates the amount of shading that occurs in each month and determines how much electricity an array could produce with that amount of shade. Credit: Solar Pathfinder.

Determining Peak Sun Hours

Like so many things these days, there are numerous online resources that will help you determine the peak sun hours at your site. One of my favorites is NASA's online database, Surface Meteorology and Solar Energy. This is my favorite website. It provides data on solar energy anywhere in the world. To access the data, however, you'll need to set up a password-accessible account. Don't fret. It only takes a few seconds.

To use this website, you need to know the latitude and longitude of your location. Latitudes and longitudes may be entered as decimal degrees (example: 33.5) or as degrees and minutes, separated by a space (example: 33 30).

Once you have entered this information, scroll down to the first box entitled "Parameters for Tilted Solar Panels." In this box, you can click on one of several options. Click on "Radiation on Equator-Pointed Tilted Surfaces." You'll be presented with the data you requested, in this case, Monthly Averaged Radiation Incident On An Equator-Pointed Tilted Surface in kWh/m²/day which is, of course, peak sun hours. This table lists the monthly peak sun hours at different array tilt angles. It also shows the optimum tilt angle for each month and the peak sun hours at the monthly optimum tilt angles. Tables can be cut and pasted into documents for reports or for your own use later on.

The amount of sunlight on the proposed array can then used to adjust the size of the array. If the program indicates 10% shading, the array would need to be 10% larger. Alternatively, trees or branches could be trimmed to eliminate shading. If shading is severe, an alternative location would be required. For optimum performance, PV systems generally require a site where an array is unshaded 90% to 95% of the time. Anything less, in my view, is a bad compromise.

The Solar Pathfinder costs about $300. The Solar Pathfinder Assistant software costs another $190—so the total cost is just shy of $500 (as of February 2016). As you might expect, there is at least one iPhone app for shade analysis. It's called EpiShade. I haven't used it, so can't vouch for its accuracy.

Sizing an Off-grid System

As a rule, off-grid systems are sized according to the month of the year with the highest demand. In the Northern Hemisphere, in all but southern locations like Florida or Louisiana, that month is January. It is also the month that offers the least sunshine. After determining the month with the highest demand, look up its average peak sun hours during that month. (This is available on the NASA site discussed in the sidebar.) Then divide the average daily consumption by the average peak sun hours to determine the size of the array. Don't forget to adjust for efficiency and shading. For battery-based systems, a 30% efficiency derate is a good idea.

Once you know the size of your array, you must size the battery bank. Although I'll discuss battery bank sizing in detail in Chapter 7, it's important to note that battery banks are sized to provide sufficient electricity to meet a family's or business' needs during cloudy periods. We call these "battery days." Most battery banks are based on a three-to-five day reserve so they'll provide enough electricity for three to five days of cloudy weather. In sunny Colorado, I'd size a system for three battery days. In cloudier Ohio, Indiana, or Illinois, I'd size a system for five battery days. Battery banks are sized to prevent deep discharging during this period—that is, so the batteries do not discharge more than 50% of their capacity. As you will see in Chapter 7, repeated deep discharging can seriously damage batteries and will shorten their lifespan.

Sizing a Grid-connected System with Battery Backup

PV arrays in grid-connected systems with battery backup are sized much like grid-connected systems. That is, the size of the array is based on average daily demand. In these systems, however, you'll also need to size a battery bank. In this case, battery banks are sized according to the electrical requirements of *critical loads*—that is, the electrical load that would be needed during a power outage. Critical loads, as noted earlier, are usually restricted to pumps or fans of heating or cooling systems, well pumps, sump pumps, refrigerators, freezers, and a few lights in critical areas, such as kitchens. Because these loads require less energy than the entire home, battery banks are typically much smaller than in off-grid systems. In fact, they are about one-third of the size. A typical home powered by a grid-tied system with battery backup will require eight 6-volt deep-cycle batteries made for renewable energy systems wired in series. This results in a 48-volt battery bank. (More on this in Chapter 7.)

Does a Solar Electric System Make Economic Sense?

At least four options are available to analyze the economic cost and benefits of a solar electric system: (1) a comparison of the cost of electricity from the solar electric system with conventional power, (2) simple payback, (3) simple return on investment, and (4) a more sophisticated economic analysis tool known as discounting. I will discuss all four in this section, but suggest that simple payback never be used, for reasons you soon shall discover.

Cost of Electricity Comparison

One of the simplest ways of analyzing the economic performance of a solar system is to compare the cost of electricity produced by a PV system to the cost of electricity from a conventional source such as the local utility. This is a five-step process.

The first step is to determine the average monthly electrical consumption of your home or business, preferably after incorporating conservation and efficiency measures. Second, calculate the size of the system you'll need to install to meet your needs. Third, calculate the cost of the system. (A solar provider can help you with this.) Fourth, after determining the cost of the system, calculate the output of the system over a 30-year period, the expected life of a PV system. Fifth, estimate the cost per kilowatt-hour by dividing the cost of the PV system by the total output over 30 years.

Suppose you live in sunny western Colorado and are interested in installing a grid-connected solar electric system that will meet 100% of your electric needs. Your superefficient home requires, on average, 500 kWh of electricity per month, or 6,000 kWh per year. That's 16.4 kWh per day. Peak sun hours in your area is 6. To size the system, divide the electrical demand (16.4 kWh per day) by the peak sun hours. The result is 2.7 kW. Adjusting for 78% efficiency, the system should be 3.46 kWh. Let's round up to 3.5 kW. For this calculation, let's assume that the system is not shaded at all during the year.

Your local solar installer says she can install the system for $3.50 per watt (just to show you how much prices have dropped, in the previous edition published in 2009, the cost was $8/watt.) Based on a cost of $3.50 per watt, this system will cost $12,250. (In 2009, it would have cost over $27,000.) Next, subtract the 30% tax credit from the federal government from the cost of the system. The federal tax credit is based on the cost of the system including installation ($12,250) minus state or utility rebates (if any). In this case, let's assume no state or utility rebates. Thirty percent of $12,250 equals $3,675. Total system cost after subtracting this incentive is $8,575.

According to your calculations or the calculations provided by the solar installer, this system will produce, on average, 6,000 kWh per year. If the system lasts for 30 years, it will produce 180,000 kWh.

To calculate the cost per kilowatt-hour, divide the system cost ($8,575) by the output (180,000 kWh). In this case, your electricity will cost slightly less than

4.7 cents per kWh. Considering that the going rate in Colorado is currently over 10 cents per kWh, the PV system represents a pretty good investment. Bear in mind, too, that utility costs have been on a steady incline for many years.

The cost of electricity calculation is simple by design. Economists reading this analysis will surely object to its simplicity. They'll note that these calculations do not take into account important economic factors such as the cost of borrowing money to purchase a PV system. Interest payments will surely add to the cost of electricity produced by the system. For those who self-finance by taking money out of savings, this economic analysis fails to take into account lost income from interest-bearing accounts homeowners raided to pay for their system. It also fails to take into account system maintenance, insurance, or property taxes, if any. That said, some states exempt PV systems from property tax. Maintenance on grid-tied systems is typically nothing, although a new inverter will very likely be required in 10 to 15 years if you install a string inverter. Microinverters are warranted for 25 years. Insurance on a system is usually insignificant.

There could be additional costs to factor in. When connecting a PV system to the grid, utilities install a bidirectional meter or a second electric meter to track the flow of electricity to and from the grid. The meter may cost $100 to $200 dollars, maybe even more in some states.

This simple economic tool also fails to take into account the rising cost of electricity. Nationwide, electric rates have increased on average about 4.4% per year over the past 35 years. In some areas, the rate of increase has been double that amount.

Although a comparison of the cost of solar electricity to the cost of utility power ignores key economic factors, the rising cost of electricity from conventional sources will, in all likelihood, offset the cost of financing a system or lost interest. So, the analysis is not as flawed as one might think.

When calculating the cost of electricity from a solar electric system, be sure to subtract financial incentives from utilities and federal, state, and local government. In the past, financial incentives have been quite substantial. For a short period, the two publicly owned utilities in Missouri (Ameren and Kansas City Power and Light) gave a $2/watt rebate for solar systems, most of which were costing about $4/watt. Add the 30% federal tax credit to that, and the homeowner was paying only 25% of the system cost. California and Colorado also once offered generous incentives, but these and many other financial incentives have either been winnowed down to a fraction of their former selves or have been phased out entirely now that solar electricity has reached priced parity with conventional power—that is, it competes economically with utility power in many areas. (The purpose of these incentives was to kickstart the solar electric industry, which they did.)

The federal government continues to offers a generous financial incentive to those who install PV systems. Their incentive is a 30% tax credit to homeowners and businesses. That tax credit applies to systems installed before the end of 2019. A 26% federal credit applies to systems from that date until January 1, 2021. A 22% tax credit applies from that date until January 1, 2022. To learn more about incentives in your state as well as the federal tax incentives, log on to dsireusa.org. That's the Database on State Incentives for Renewable Energy. Type in your zip code, then click enter. Look for the Residential Renewable Energy Tax Credit, though there may be other incentives to check out as well.

If you own a business, you can receive even greater financial incentives. The feds, for instance, allow business owners to depreciate a solar electric system on an accelerated schedule, which means they can deduct the costs in five years, faster than other business equipment. This further reduces the cost of a PV system.

Sweetening the pie even more, the US Department of Agriculture offers a 25% grant to cover the cost of PV systems on farms and businesses in rural areas. Rural businesses means any business in an area with a population under 50,000, so long as they are not owned by a large corporation. That includes hair salons, restaurants, auto repair, antique shops, copy shops, etc. The USDA's minimum grant is $2,500 (for a $10,000 system) and the maximum is $500,000. A business that avails itself of the 30% federal tax credit is also able to depreciate the system over five years; if it also receives a USDA grant, a system can get dirt cheap. Let's do the math. Suppose a 10-kW PV system cost $35,000. The USDA grant would decrease the cost to $26,250. The federal tax credit would decrease the cost to $18,375. Accelerated depreciation would lower the cost even further, depending on the business' tax rate.

Clearly, financial incentives can dramatically reduce the cost of a PV system. Consequently, many PV system installations are driven by incentives. Incentives have been a great marketing tool, as well.

When comparing the cost of an off-grid PV system on a new home to the cost of electricity from the grid, don't forget to include the line extension fees—the cost of connecting the home to the electrical grid. If your home or business is more than a few tenths of a mile from existing electric lines, you could be charged handsomely to run lines to your home. I've seen clients pay $20,000 to run a line 0.2 miles (0.32 km) and $65,000 for a one-mile (1.6 km) line run. Remember that the cost of connecting to the electric grid only covers the cost of the installation of poles, electric lines, and a meter. It does not buy you a single kilowatt-hour of electricity.

Calculating Simple Return on Investment

Another relatively simple method used to determine the cost-effectiveness of a PV system is *simple return on investment* (ROI). Simple return on investment is, as its name implies, the savings generated by installing a PV system. It is expressed as a percentage.

Simple ROI is calculated by dividing the annual dollar value of the energy generated by a PV system by its cost. A solar electric system that produces 6,000 kWh of electricity per year at 10 cents per kilowatt-hour generates $600 worth of electricity each year. If the system costs $8,575, after rebates, the simple return on investment is $600 divided by $8,575 × 100 which equals 6.99%. Let's call it 7%, If the utility charges 15 cents per kWh, the 6,000 kWh of electricity would be worth $900, and the simple ROI would be 10.5%. Both of these represent very decent rates of return. Consider what your money would be doing otherwise. In the States, most banks are pay 0.125% interest. Some national online savings accounts offer a whopping 0.95% at this writing.

As in the cost comparison method, simple ROI fails to take into account interest payments on loans required to purchase the system or opportunity costs, for example, lost interest if the system is paid for in cash. It also fails to take into account system maintenance, insurance, or property taxes, if any. All of these factors could decrease the return on investment.

But, as in the cost of electricity comparison, this method fails to take into account a number of factors on the other side of the ledger—for example, rising electricity costs. Simple ROI also fails to take into account the fact that money saved on the utility bill is tax-free income to you. There are also possible income tax benefits for businesses, such as accelerated depreciation.

Despite these shortcomings, simple return on investment is a convenient tool for evaluating the economic performance of a renewable energy system. It's infinitely better than the black sheep of the economic tools, *payback* (also known as "simple payback").

Why?

Payback is a term that gained popularity in the 1970s. It was used to determine whether energy conservation measures and renewable energy systems made economic sense. Payback is the number of years it takes a renewable energy system or energy-efficiency measure to pay back its cost through the savings it generates.

Payback is calculated by dividing the cost of a system by the anticipated annual savings. If the $8,575 PV system we've been looking at produces 6,000 kilowatt-hours per year and grid power costs you 10 cents per kWh, the annual savings of

$600 yields a payback of 14.3 years ($8,575 divided by $600 = 14.3). In other words, this system will take almost 14.3 years to pay for itself. From that point on, the system produces free electricity.

While the payback of 14.3 years seems ridiculously long, don't forget that this is the system that yielded a simple very respectable 7% return on investment, which looked pretty darn good.

While simple payback is popular, this example shows that it has very serious drawbacks. The most important is that payback is a foreign concept to most of us and, as a result, can be a bit misleading.

Besides being misleading, simple payback is a concept we rarely apply in our lives. Do avid anglers ever calculate the payback on their new bass boats? ($35,000 plus the cost of oil, gas, and transportation to and from favorite fishing spots divided by the total number of pounds of edible bass meat at $5 per pound over the lifetime of the boat?) Do couples ever calculate the payback on their new SUV or new chandelier in the dining room?

Interestingly, simple payback and simple return on investment are closely related metrics. In fact, ROI is the reciprocal of payback. That is, ROI = 1/payback. Thus, a PV system with a 10-year payback represents a 10% return on investment (ROI = $\frac{1}{10}$). A PV system with a 20-year payback represents a 5% ROI.

Although payback and ROI are related, return on investment is a much more familiar concept—and a more useful one. We receive interest on savings accounts and are paid a percentage on mutual funds and bonds—both of which are a return on our investment. Many of us were introduced to return on investment very early in life—when we opened our first interest-bearing savings account. Renewable energy systems also yield a return on our investment, so it is logical to use ROI to assess its economic performance.

Net Present Value: Comparing Discounted Costs

For those who want a more sophisticated tool to determine whether an investment in solar energy makes sense, economists have devised an ingenious though somewhat complicated method that allows us to compare the value of a solar electric system to the cost of buying electricity from a utility for the next 30 years. It allows us to make this rather unusual comparison based on the value of today's dollar (in the United States and Canada)—or whatever currency your country uses. They call the value of something in today's dollar *present value*.

Unlike the previous methods, present value takes into account numerous additional economic factors besides the cost of the system. These can include

Table 4.4. Economic Analysis of PV System in Colorado

Year	Discount Factor 3.0%	Buy Utility Electricity Cost 4.4%	Discounted Cost	Proposed PV System Cost	Discounted Cost
0	1.000	$0	$0	$9,555	$10,710
1	0.971	$570	$553	$0	$0
2	0.943	$595	$561	$0	$0
3	0.915	$621	$569	$0	$0
4	0.888	$649	$576	$0	$0
5	0.863	$677	$584	$0	$0
6	0.837	$707	$592	$0	$0
7	0.813	$738	$600	$0	$0
8	0.789	$771	$608	$0	$0
9	0.766	$804	$617	$0	$0
10	0.744	$840	$625	$0	$0
11	0.722	$877	$633	$0	$0
12	0.701	$915	$642	$0	$0
13	0.681	$956	$651	$0	$0
14	0.661	$998	$660	$0	$0
15	0.642	$1,042	$669	$0	$0
16	0.623	$1,087	$678	$0	$0
17	0.605	$1,135	$687	$0	$0
18	0.587	$1,185	$696	$0	$0
19	0.570	$1,237	$706	$0	$0
20	0.554	$1,292	$715	$3,200	$1,772
21	0.538	$1,349	$725	$0	$0
22	0.522	$1,408	$735	$0	$0
23	0.507	$1,470	$745	$0	$0
24	0.492	$1,535	$755	$0	$0
25	0.478	$1,602	$765	$0	$0
26	0.464	$1,673	$776	$0	$0
27	0.450	$1,746	$786	$0	$0
28	0.437	$1,823	$797	$0	$0
29	0.424	$1,903	$808	$0	$0
30	0.412	$1,987	$819	$0	$0
Total		$34,191	$20,330	$12,755	$9,555

maintenance costs, insurance, inflation, and the rising cost of grid power. As all readers know, inflation decreases the value of money over time. Economists refer to this as the *time value of money*. The time value of money takes into account the fact that a dollar today will be worth more than a dollar tomorrow and even more than a dollar a year from now. Economists refer to rate at which the time value of money declines as *discount factor*.

To make life easier, this economic analysis can be performed by using a spread sheet like the one shown in Table 4.4 provided by renewable energy economist John Richter of the Sustainable Energy Education Institute in Michigan. This spread sheet is available for your use by emailing me at danchiras@ever greeninstitute.org.

To understand how this system works, take a look at Table 4.4. The first column is the year. Note that it runs from 1 to 30—that's 30 years of service. The second column includes the discount factor. I used the rate of inflation. Using this number, a dollar today will be worth 41 cents in 30 years (see column 2).

The next column (under the heading "Buy Utility Electricity") shows the cost of electricity from the local power company—that is, how much you will pay each year for electricity purchased from the local utility—taking into account rising fuel costs (4.4% annual increase). As shown here, the PV system produces $570 worth of electricity in year one. If you were to buy that electricity from the utility instead, it would cost you $570 in year one and $595 in year two. You are still buying 6,000 kWh electricity. It's just going to cost you $25 more because the utility raised its rates by 4.4%. In ten years, your checks to the utility company for your 6,000 kWh of electricity will total $840. The last entry in this column is the total cost of electricity to you—$34,191. That's how much money you will pay the utility over a 30-year period if you purchase 6,000 kWh of electricity per year from them, based on an inflationary increase of 4.4% per annum. (It's the sum total of all your checks or automatic withdrawals.)

The next column under the heading "Buy Utility Electricity" is the discounted cost of electricity from the utility. The discounted cost of electricity from the utility is the cost of electricity taking into account the discount rate (the declining value of the dollar due to inflation) applied to the rising cost of electricity.

Take a look at year 10. As you can see, you will pay the utility $840. In today's dollars, those checks would be worth $625.

In the bottom of third column is the sum of all the checks you have written to the utility—or automatic withdrawals you've authorized. It comes to $34,191. The next column shows you the value of the money in today's dollars (its present value). It comes to $20,330.

In the fifth column of the spreadsheet is the cost of the PV system—$9,555. You will note that I added a $3,200 charge after 15 years to replace the inverter. As shown at the bottom of this column, over a period of 30 years, you will have invested $12,755 in your system. That's the sum of all the checks you have written—all two of them.

The last column of the spreadsheet is the discounted cost of the PV system. This is the present value of your expenditure, taking into account the discount factor of 3%. In other words, it adjusts your expenditures for inflation. This number comes to $9,555.

The final step is to compare the present value of the electricity purchased from the utility ($20,330) to the present value of the system ($9,555). In this example, the difference between the two is *net present value* of the PV system: $10,775. In other words, investing in solar is over $10,000 less expensive than simply paying the utility for 30 years.

In this type of analysis, if a solar electric system is cheaper than buying electricity, it makes economic sense. If it costs more, it doesn't. The greater the difference in the cost of the two systems, the more compelling the decision. Even if the differential is small, however, the investment may be worth it because solar electricity offers so many other benefits, discussed shortly.

Comparisons based on net present value, return on investment, or the costs of electricity are vital to making a rational decision about a PV system. As energy economist John Richter pointed out to me, "Even though a system may not make perfect sense from an economic standpoint, it is your money. You can spend it how you see fit." You may want to purchase a PV system for peace of mind—knowing it will free you from utility power and rising fuel costs. Being independent of the local power company is often a compelling motivation. The idea of selling power to the utility may motivate others (even if this is often an illusion—see sidebar, "Selling Energy to the Utility—Fact or Fiction?"). Or, you may find the personal satisfaction

Selling Energy to the Utility—Fact or Fiction?

Many people are enamored of the idea of selling surplus electricity to the utility. Some even imagine that they'll get rich. Is this realistic?

In the previous example of a system that provides 100% of the owner's 500 kWh per month load, the system cost $8,575 after rebates. At 10 cents per kWh, this homeowner is saving $600 per year. How much is he selling back to the utility?

Nothing.

This system was sized to meet 100% of his demand, so there is nothing left over to sell to the utility. What if the size—and cost—of the system were doubled? Then the owner would produce 500 kWh per month excess energy to sell to the utility. What will the utility pay for this energy?

That depends.

Let's say the customer's utility pays the retail rate—that is, the amount it charges its customers for electricity they purchase from the utility. The 500 kW of electricity would be worth about $50 per month, or $600 per year (if the retail cost of electricity is 10 cents per kWh). Not a bad investment.

Reimbursement for surplus at retail rates occurs in many states, including Colorado, Kansas, Kentucky, and Minnesota. Unfortunately, many states reimburse at a much lower rate, known as avoided cost. This is the utility's wholesale cost for producing power, which is far less than the retail price it charges its customers. Many states have adopted this pricing in their net metering policies. The avoided cost is usually about one-fourth of the retail rate. So, if you are buying electricity at 10 cents per kWh, you are being reimbursed at 2.5 cents kWh for your surplus. There have been months when I have generated 900 kilowatt-hours of electricity and been reimbursed about $23. The utility sold that same electricity to my neighbors for $63. Reimbursement rates vary from state to state, but all reimburse at either wholesale or retail. Be sure to check your state's reimbursement policy (see net metering policies on the dsireusa.org website) before installing a grid-connected PV system.

If your state reimburses at retail rate, you could make some money if you oversize your system. Will you get rich? No, but a 7% return on your investment isn't bad given the prevailing interest rates in today's economy. If your state requires utilities to reimburse at retail rates and you double the size of your system, you will be generating all of your household electricity at a much lower cost than the utility and have a modest stream of income that will last for many years. That said, it is important to point out that state laws sometimes limit the size of customer PV systems to the amount of electricity a household normally consumes.

in generating power from a clean, renewable resource sufficient enough reason to invest in a PV system. Or, for some of us, it may be enough just to have a fancy new toy to play with.

Alternative Financing for PV Systems

Many individuals and businesses do not have the financial wherewithal to purchase a PV system outright, even with state, local, or federal incentives. To purchase systems, some people take out loans—home equity loans or small business loans—to finance their systems. Unfortunately, not everyone wants to incur that kind of debt. If you are one of these people, there are some alternative financing mechanisms that could make your dreams of a PV system come true: *power purchase agreements* (PPAs) and *leases*.

PPAs and leases are a pretty painless way to get into solar, which is why they are so popular. According to recent estimates, about two-thirds of all solar electric systems are currently being installed under one of these types of arrangement.

PPAs and leases vary from one company to the next. Within the same company they often share some similarities, so these options can get pretty confusing. In both PPAs and leases, for instance, the companies install the systems for the customer, usually at no cost. Some companies may offer the homeowner an option to invest in the system in trade for lower electrical bills. Companies typically agree to repair any damage they might cause when installing a system, especially roof damage. The company, in turn, owns, monitors, insures, and maintains the system for the duration of the agreement.

PPAs and leases are long-term agreements, typically 20 to 30 years. At the end of the lease or PPA, customers can renew their agreement or enter into a new agreement, which often entitles them to newer equipment. If a customer wishes to terminate his or her agreement, the companies will remove the solar systems and restore the roof so it won't leak—at no cost to the customer. (Be sure you study the contractual details on this provision.) What if a customer sells his or her home? Check the company's contract carefully, but leases and PPAs are typically fully transferable should a homeowner decide to sell. If the new buyer is interested, he or she can take advantage of the solar system on the roof.

Both PPA and lease options are structured to allow homeowners to make monthly payments, which helps make solar electricity a lot more affordable. In PPAs, customers pay for the electricity they consume at a fixed rate per kilowatt-hour. (You are purchasing power.) In leases, however, customers pay a fixed monthly amount (kind of like budget payments made to electric companies). To determine this cost, the company simply calculates how much electricity the system should produce each year, divides it by 12, then bills you monthly for $\frac{1}{12}$ of the annual total, often at a rate that is slightly below what the utility is charging. What if you need more electricity than your PV system generates? Lease and PPA PV systems are grid-tied. If you need more electricity, you purchase it from your local utility. Check this carefully, as numerous utilities are now charging their solar customers a premium for electricity they purchase from them. At the end of the month, then, expect two bills—one from the solar leasing company and one from the utility.

What if the solar system underperforms?

In PPAs, homeowners simply purchase more utility power. However, it's not always that simple. In lease agreement with SolarCity, for example, if their PV system underperforms, the company reimburses its customers for the difference. (Be sure to check out the details on this very carefully.)

PPAs and leases are structured to make the companies that offer them a respectable profit. Nothing wrong with that. To do so, these companies reap all the tax incentives, rebates, and accelerated depreciation. They also purchase modules and equipment in bulk—by the tractor trailer load—so they can install systems inexpensively.

PPAs and leases have allowed tens of thousands of homeowners to generate their own electricity from solar energy. According to their website, SolarCity not only has tens of thousands of homeowner customers, they have installed PV systems on "more than 400 schools, including Stanford University, government agencies such as the U.S. Armed Forces and Department of Homeland Security, and well-known corporate clients, including eBay, HP, Intel, Walgreens and Walmart."

Although power purchase agreements have traditionally been used to bring PVs to single family homes, they have also been used to finance rooftop PV systems for entire real estate developments in California, thanks to the work of Open Energy Corporation. This company finances the entire project so there's no upfront cost to the developers or future homeowners. The benefits to developers are several. Builders receive the tax credits and enjoy robust sales—their units sell four times faster than comparable, nonsolar residences in the area. Homeowners benefit, too. Not only do they incur no upfront costs, but they enjoy lower electric bills and live a more environmentally friendly lifestyle. They also own a residence that will sell more quickly and at a higher price when the time comes to put it on the market.

The success of low-entry-cost PV systems has been due to private funding. SolarCity, for example, has teamed up with investors like Morgan Stanley, Google, Honda, USBankcorp, and host of others to provide leases for 217,000 systems in

19 different states, as of this writing (February 2016). (They provide funding to the company's lease program.)

Customers who lease PV systems not only end up paying less than they would if they were to purchase power from the utility, they pay a fixed amount per month, which often makes it easier to budget. So should you go with a lease or a PPA?

Representatives from the industry admit that the financial costs (to the customer) are not that different over the long haul. Bottom line, solar leases and PPAs make it possible for any homeowner or business owner to stop talking about solar and start generating electricity from the Sun.

If you'd like to power your home with solar electricity, but can't afford a system or don't want to borrow the money, consider a lease or a power purchase agreement. If you can afford a system of your own, and receive generous rebates, you may want to consider installing one yourself. The economic benefits can be many and can result—under optimum conditions—in very affordable and clean electricity.

Power purchase agreements (PPAs) and lease agreements vary from one company to the next, so be sure to shop around and be sure to study your options very carefully. Pay special attention to the details on monthly payments in the contract. Look for provisions that allow the company to increase its rates or flat-rate payments. In other words, if you initially are buying electricity at a certain rate, say 15 cents per kilowatt-hour, can the company increase the rate they charge for electricity over time to adjust for inflation? If you will be locked into a monthly payment schedule, say $70 per month 12 months a year, will that increase during the lease, and by how much? Be sure to talk to other customers and check out complaints online. There are plenty of them. Don't rely solely on the very convincing websites.

Community Solar

If you lack the financial ability to buy a solar system outright or if you rent a home or apartment and can't install a system on your landlord's roof, or if you live in a home that's heavily shaded, there's yet another option: *community solar*.

Community solar started as a private venture in California, then spread to Colorado, two of the most solar-friendly states in the nation. In these early instances, the companies installed the large arrays on solar farms or the roofs of warehouses. The companies then sold modules (from one to several dozen) to subscribers in the city or town in which they were located, based on individual customer demands.

Here's what's so cool about community solar: With the full cooperation of local utilities, the monthly electrical production of an individual's modules is credited to his or her utility bill.

This concept, originally called *remote net metering* has spread to numerous states, including Massachusetts, Minnesota, and Georgia. Be sure to check this option out. For more information, including case studies, be sure to check out Northwest Community Energy's website: nwcommunityenergy.org/solar.

Solarize Campaigns

Community solar is a great idea, as are solar leases. However, there are other innovative ways to help make solar more affordable. If you would like a system on your home, you could possibly tap into group buying power. That is, you may be able to capitalize on volume discounts acquired by nonprofit organizations, local solar installers, and even some local governments.

One of the first projects of this nature took place—where else?—in Portland, Oregon, in 2008. Working with a nonprofit neighborhood group and the Energy Trust of Oregon, the citizens banded together to purchase solar equipment in bulk. They received a volume discount that reduced the cost of their systems by 30%.

Similar projects have been undertaken in nearly 20 states in United States. For more information, check out the US Department of Energy's *The Solarize Guidebook*. Also check out Vaughan Woodruff's piece on the topic in *Home Power*, Issue 171.

Putting It All Together

In this chapter, you've seen that efficiency measures lower the size and cost of a system, often saving huge sums of money. Tax incentives and rebates also lower upfront costs. Some states exempt PV systems from sales taxes or property taxes, creating additional savings. Avoiding line extension fees by installing an off-grid system in a new home rather than a grid-connected system can also save huge amounts of money, often enough to pay for a good portion, or perhaps even all of your system cost. If buying a system isn't possible, you have several options: leases, power purchase agreements, community solar, and solar co-ops.

I encourage folks who are building superefficient passive solar/solar electric homes to view savings they'll accrue from efficiency measures and passive solar design as a kind of internal subsidy for their PV systems. My off-grid solar electric system in Colorado cost about $17,000 and generated $4,000 worth of electricity in the first 12 years. The return on investment was pretty low. However, this passive solar home has saved me approximate $15,000 in heating bills during this same period. Savings on electricity from the PV and savings on heating bills resulting from passive solar heating have more than paid for my system.

I accrued additional savings by avoiding the line extension fee of $2,000. Avoiding the $20/month meter-reading fee the local utility wanted to charge saves me another $240 per year or nearly $3,400 in the first 14 years. Even though I had to replace my batteries after 11 years, which cost $3,400, I was clearly ahead. What's the balance?

System cost, including the new batteries: $20,400.

Systems savings: $24,400.

Economics is where the rubber meets the road. Comparing solar electric systems against the "competition," calculating the return on investment, or comparing strategies using present value, gives a potential buyer a much more realistic view of the feasibility of solar energy. Just don't forget to think about *all* the opportunities to save money. If you invest in efficiency measures to lower the system cost, remember that those efficiency measures will provide a lifetime of savings, helping to underwrite your PV system. As I pointed out in Chapter 1 and again in this chapter, economics is not the only metric on which we base our decisions. Energy independence, environmental values, reliability, the cool factor, bragging rights, the fun value, and other factors all play prominently in our decisions to invest in renewable energy.

People often invest in renewable energy because they want to do the right thing. If you want to invest $12,000 to $40,000 in a PV system to lower your carbon footprint, create a better world for your children or grandchildren, or simply to live by your values, do it. It's *your* life. It's *your* money.

CHAPTER 5

Solar Electric Systems:
What Are Your Options?

The first decision you will need to make when going solar is what type of system suits your needs. As noted in Chapter 4, solar electric systems fall into three categories: (1) grid-connected, (2) grid-connected with battery backup, and (3) off-grid.

In this chapter, I'll examine each system and its main components. I'll discuss the pros and cons of each system. Plus, I'll explore hybrid renewable energy systems—those that couple PV electric systems with other renewable energy technologies such as wind energy.

Grid-connected PV Systems

Grid-connected PV systems are by far the most popular solar electric system on the market today. As shown in Figure 5.1, grid-connected systems are so named because they are connected directly to the electrical grid—the electric wires that supply energy to your community. These systems are also referred to as *batteryless grid-connected* or *batteryless utility-tied* systems because they do not employ batteries to store surplus electricity.

In grid-connected systems, excess electricity produced by PV systems is fed onto the local electrical distribution system—the wires that run by our homes and businesses. These wires, however, also provide power when homes or businesses supplied by solar electricity need more than their systems are producing or at night when their systems are dormant.

As shown in Figure 5.3, a grid-connected system consists of five main components: (1) a PV array, (2) an inverter designed specifically for grid connection, (3) the main service panel (or breaker box in very old homes not wired to modern code requirements), (4) safety disconnects, and (5) the utility meter. Safety disconnects include the AC and DC disconnect. The system may also contain a DC combiner box.

To understand how a batteryless grid-connected system works, let's begin with the PV array. The PV array produces DC electricity. This flows through wires to the *inverter*, which converts DC electricity to AC electricity. This "central" inverter is typically wired to one to four series strings of modules. For this reason, it is technically referred to as a *string inverter*. (For a description of AC and DC electricity, see the sidebar "AC vs. DC Electricity.")

String inverters may be mounted indoors or outdoors, for example, on the exterior wall of a home or on the rack of a ground-mounted PV array. As you shall soon see, I think indoors is the best option.

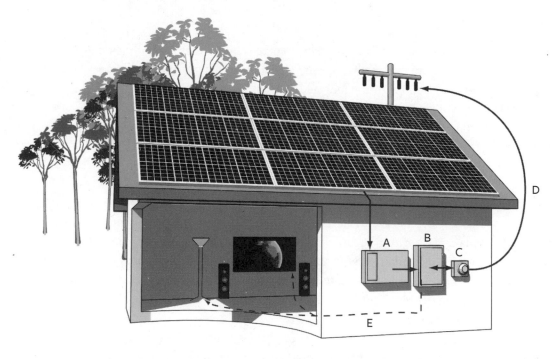

FIGURE 5.1. Grid-connected PV System. Diagram of a simplified batteryless grid-tied PV array. Systems such as this usually consist of two or more series strings of modules. Systems are typically wired at 350 to 450 V DC. This system is Code compliant. Grid-tied inverters come with a combined AC/DC disconnect that typically meets all the requirements of the National Electric Code. Some installers wire series strings into a combiner box, which then feeds to a DC disconnect located between the array and the inverter. Utilities commonly require an AC disconnect between the main service panel and the utility meter. Credit: Anil Rao.

Grid-tied or Utility-tied: What's in a Name?

Technically, utility-tied is the most accurate name for a grid-tied system, although few installers use it. The reason this name is more accurate is because the term "grid" refers to the high-voltage electrical transmission system that crisscross countries (Figure 5.2). They feed electricity into local electrical distribution systems in cities and towns often owned by utilities. Customers are actually connected to the grid through their local utility's distribution system, as shown in Figure 5.2b, so the terms "utility-connected" or "utility-tied" are more accurate than "grid-connected" and "grid-tied." However, since it is used most often, we'll stick with the term "grid-connected."

FIGURE 5.2. High-Voltage Electrical Wires. (a) The national electric grid consists of high-voltage wires and extremely tall towers that transmit electricity across states, allowing utilities to share electricity. Credit: Dan Chiras. (b) The national electric grid feeds into local utility networks that serve our homes and businesses. Credit: Forrest Chiras.

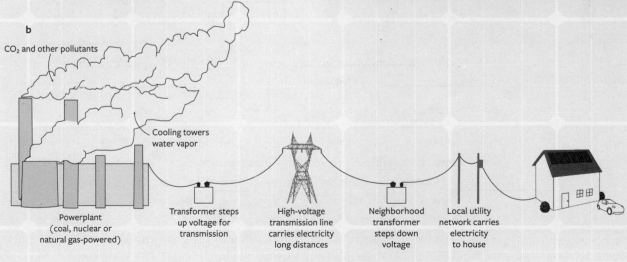

CO$_2$ and other pollutants

Cooling towers water vapor

Powerplant (coal, nuclear or natural gas-powered)

Transformer steps up voltage for transmission

High-voltage transmission line carries electricity long distances

Neighborhood transformer steps down voltage

Local utility network carries electricity to house

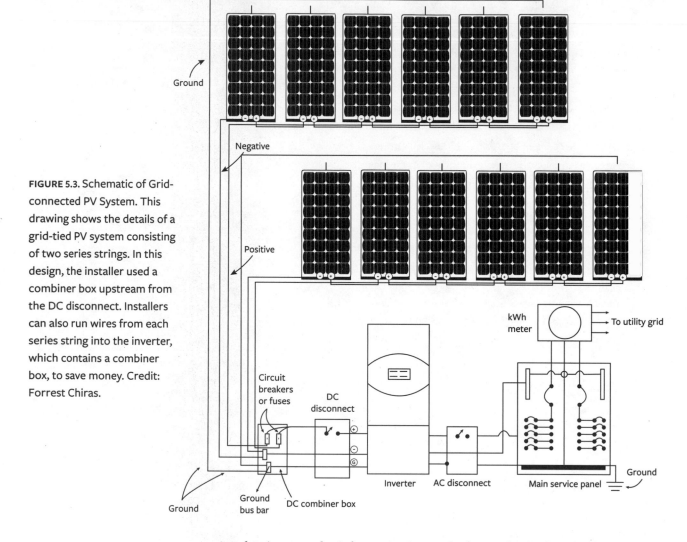

FIGURE 5.3. Schematic of Grid-connected PV System. This drawing shows the details of a grid-tied PV system consisting of two series strings. In this design, the installer used a combiner box upstream from the DC disconnect. Installers can also run wires from each series string into the inverter, which contains a combiner box, to save money. Credit: Forrest Chiras.

Another inverter that's becoming increasingly popular is the *microinverter*. As shown in Figure 5.4, microinverters are miniature inverters. They are wired directly into the modules—one microinverter per module.

Microinverters convert the 36- to 38-volt DC electricity produced by modules into 240-volt AC electricity at the module. Multiple microinverters are then wired together in series to boost the amperage. (We typically wire up to 16 microinverters in an AC series string.) Microinverters send electricity to an AC utility disconnect. From the AC disconnect, electricity flows directly into the main panel (Figure 5.5).

PV systems are wired into a main panel via a 240-volt circuit breaker. It is installed either at the top or bottom of the panel, specifically, the end of the main panel farthest from the main breaker. The National Electric Code requires that the PV system breaker be labeled so any electrician working on the panel will know where/what it is. The main panel must also be labeled to indicate that it is powered by a second source of energy, a PV system.

Microinverter-based PV systems are the simplest of all grid-tied systems to wire. They are slightly more efficient than systems with string inverters. One reason for this is that because they are mounted on and serve individual modules, microinverters reduce the negative effects of shading on an array. While shading part of an array that feeds a string inverter can reduce performance of the entire

FIGURE 5.4. Microinverter on an Array. Microinverters like the one shown here are mounted either on the module frames or the rack system in close proximity to PV modules. They convert the DC electricity into 240 V AC electricity. Credit: Dan Chiras.

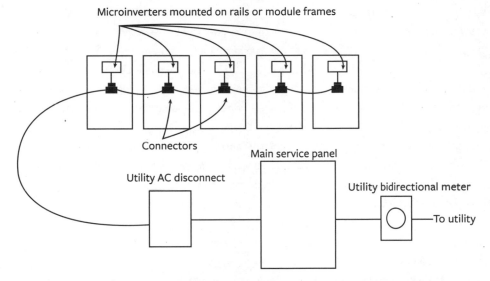

FIGURE 5.5. Schematic of Grid-connected PV System with Microinverters. This drawing illustrates the components of a grid-tied solar electric system with microinverters. In this type of array, the microinverters are wired in series, which increases the amperage of the series string. This system requires an AC disconnect, and is wired directly into the main panel. Credit: Forrest Chiras.

array, shading on one or two modules equipped with microinverters only affects the output of those modules. (In my opinion, this advantage isn't as huge as manufacturers suggest because all PV modules come with bypass diodes that divert electricity around shaded portions of modules.) A final advantage of microinverters—and this is enormous—is that they come with a 25-year warranty.

The inverter doesn't just convert the DC electricity to AC, it converts it to grid-compatible AC—that is, 60-cycles-per-second, 240-volt electricity—the kind we have in our homes. (See sidebar "Frequency and Voltage" for more.) Because the inverter produces electricity in sync with the grid, inverters in these systems are often referred to as *synchronous inverters*.

The 240-volt AC electricity produced by the inverter flows to the main service panel, a.k.a. the breaker box. The main service panel is also known as the "main

AC vs. DC Electricity

Electricity comes in two basic forms: *direct current* (DC) and *alternating current* (AC). Direct current electricity consists of electrons that flow in one direction through an electrical circuit. It's the kind of electricity produced by a flashlight battery or the electrical systems in automobiles. Even cell phones, laptop computers, and televisions are powered by DC electricity. Although we plug them into outlets that carry AC electricity, these devices contain circuitry that converts the AC to DC. In a TV, the converter is inside the unit. In a computer, it is in the power supply transformer.

DC electricity is produced by solar modules. Most other sources of electricity, including wind turbines and conventional power plants, generate alternating current electricity, which is the type of electricity we use in our homes and businesses.

Like DC electricity, AC electricity consists of the flow of electrons through a conductor. However, in alternating current, the electrons flow back and forth. That is, they change (alternate) direction in very rapid succession, hence the name "alternating current." Each change in the direction of flow (from left to right and back again) is called a cycle.

In North America, electric utilities produce electricity that cycles back and forth 60 times per second. It's referred to as 60-cycle-per-second—or 60 Hertz (Hz)—AC. The hertz unit commemorates Heinrich Hertz, the German physicist whose research on electromagnetic radiation provided a foundation for radio, television, and wireless transmission. In Europe and Asia, utilities produce 50 Hz AC.

In both AC and DC electricity, electrons flowing through a conductor (usually a wire) contain energy. In a DC circuit running directly from a PV module, the energy these electrons contain was imparted to them by solar energy. Remember, solar energy ejects the electrons from the outermost shells of the atoms in PV cells. This energy stays with the electron as it flows out of PV cell and through the wires. It is captured and used by various electric loads (devices that consume energy) to do work, for example, to run a pump or power a DC light.

Frequency and Voltage

Alternating current is characterized by a number of parameters, two of the most important being frequency and voltage. Frequency refers to the number of times electrons change direction every second and is measured as cycles per second. (One cycle occurs when the electrons switch from flowing to the right then to the left then back to the right again.) In North America, the frequency of electricity on the power grid is 60 cycles per second (also known as 60 Hertz).

The flow of electrons through an electrical wire is created by a force. Scientists refer to this mysterious electromotive force as *voltage*. The unit of measurement for voltage is volts.

Voltage is a more difficult electrical term to understand. You can think of it as electrical pressure, as it is the driving force that causes electrons to move through a conductor such as a wire. Without this force, electrons will not move through a wire. Voltage is produced by batteries in flashlights, solar electric modules, wind generators, and conventional power plants.

panel" or simply the "panel." Main panels are typically located indoors—in basements or garages. In newer homes, main panels are typically rated at 200 amps. If you have an all-electric home, though, you may have two 200-amp panels.

From here, electricity flows through the wires in a building to active loads—that is, to electrical devices that are operating. Each circuit is referred to as a *branch circuit*. If the PV system is producing more electricity than is needed to meet these demands—which is often the case on sunny days—the excess automatically flows onto the grid, or, as the experts say, the surplus is "backfed" onto the grid. The system functions in sync with the utility's system.

When a PV system is producing more electricity than is required, the surplus travels from the main service panel through the utility's electric meter, typically mounted on an exterior wall, that is, outside of the house. Utility meters may also be mounted on nearby electric poles (Figure 5.6). From the meter, electricity flows through the electric wires

FIGURE 5.6. Utility Meter Mounted on Pole. In most homes, utility meters are mounted on the side of the building near the main service entrance—where the electricity enters the home from the utility. Utility meters may also be mounted on poles near a home or business. In this system, two meters were installed, one to track electricity from the grid and the second to track solar electricity fed back onto the grid. These meters are read remotely by the utility over the electrical line. Credit: Dan Chiras.

that connect homes and business to the utility lines. From here, it travels along the power lines running underground, if you are lucky, or on poles located along the streets near our homes and businesses. This electricity supplies neighboring homes and businesses. Once the electricity is fed onto the grid, the utility treats it as if it were theirs. That is to say, your neighbors pay the utility for the electricity you generated.

The utility meter monitors a PV system's contribution to the grid and all the electricity delivered to customers from the grid. This is commonly accomplished by installing a digital meter, technically referred to as a *bidirectional meter*. In areas where older dial-type meters are still in service, the disk of a bidirectional meter spins one direction when electricity is flowing off the grid, but flows backward when electricity is flowing from the inverter onto the grid (Figure 5.7a). If a more modern digital meter is in use, the meter keeps track of electricity received and electricity backfed onto the grid electronically.

As shown in Figure 5.6, in some instances utilities install two meters—one that monitors the flow of electricity delivered from the grid and another that monitors electricity received from a PV system. (To learn how an electric utility "pays" for surpluses, check out "Net Metering and Billing in Grid-connected Systems" on page 123.)

In addition to the utility electric meter, code-compliant grid-connected solar electric systems also contain two safety disconnects. Safety disconnects are manually operated switches that enable service personnel to disconnect at key points in the system to prevent electrical shock when performing service. The National Electric Code (NEC) requires a disconnect in the DC wiring (wires from the array to the inverter) and the AC wiring (wires running from the inverter to the main panel). All grid-tied inverters have a built-in AC/DC disconnect that meets this requirement. The NEC requires that this switch be labeled as such.

Even though the NEC requirement is met by the inverter-based AC/DC disconnect, there are times when an additional DC disconnect is installed (or required), for instance, if the array is mounted some distance from the inverter (Figure 5.8). In one of my systems, the PV array is about 250 feet from the house, where the inverter's located. I installed a DC disconnect at the array to make it easier to service the array. It comes in handy. There's no need to walk all the way to the house to turn the array on and off. The NEC requires that this switch be properly labeled.

In most states, grid-connected systems also contain a manual AC disconnect. This is known as a *utility disconnect*. That's because it allows utility workers to isolate or disconnect a PV system from the electrical grid, should they need to work on an

a

b

FIGURE 5.7. (a) Dial-type and (b) Digital Electric Meters. These photos illustrate the two types of utility meter commonly encountered in solar installations. In older homes, dial-type analog electric meters are frequently found; utilities now usually switch them out for digital electric meters. Credit: Dan Chiras.

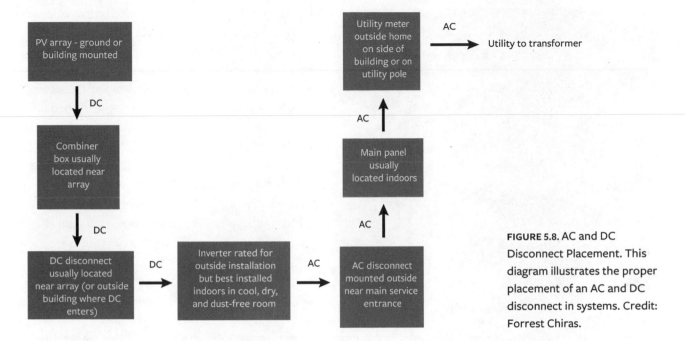

FIGURE 5.8. AC and DC Disconnect Placement. This diagram illustrates the proper placement of an AC and DC disconnect in systems. Credit: Forrest Chiras.

electric line. It's designed to protect the utility workers from shock caused by electricity backfed onto the grid by a PV system. If an electric line is downed in a storm, for instance, utility workers can be sure that they won't be shocked or electrocuted by electricity still being backfed onto the grid.

Shown in Figure 5.9, the utility AC disconnect must be mounted outside, typically in close proximity to the building's utility service entrance (where electrical wires enter a building). This switch must accessible, visible, and lockable. Accessible means that the switch is easy for utility workers to locate and easy to reach. They have the legal right to shut a customer off from the grid should they need to work on the line. For ease of access, utilities often require that it be located within five feet of the main service entrance where the utility meter is located.

The utility disconnect must also be visible. This means that it must be a switch that one can visually inspect to be sure the circuit has been opened. This is achieved by a knife-blade switch. The blades are operated by the switch handle. When in the off position, it's clear that the blades are no longer in contact with the slots into which they fit, creating an open circuit that cannot conduct electricity. Circuit breakers cannot be used for an AC disconnect (or a DC disconnect, either). Although circuit breakers can open and close a circuit, a circuit breaker may malfunction, and there is no way to visually confirm that the circuit is truly open.

The utility disconnect must also be lockable. As shown in Figure 5.9, the handle on the AC connect can be locked in the off position by the utility by inserting a lock through the hole. Locking prevents a member of a household from switching the service on while utility workers are repairing a downed line or addressing some other issue. (A homeowner cannot install a lock on this switch.)

The NEC requires that all AC disconnects be clearly labeled "Utility Disconnect" or "Interconnection Disconnect Switch" using a label that will withstand all weather conditions and UV light (Figure 5.10).

For many years, lockable AC disconnects were considered critical for the safety of utility personnel. That's because the 240-volt electricity backfed onto the grid from an inverter is boosted to between 12,000 to 24,000 volts by the utility transformer. That's the voltage of the electric lines the power companies use to deliver electricity to a building.

Although accessible, lockable, visible AC disconnects are required by most utilities, large utilities in California and Colorado—that have thousands of

FIGURE 5.9. AC Disconnect Showing Details. This photograph shows the internal workings of an AC, or utility, disconnect. Notice the knife blades that are operated by the handle. This disconnect must also be lockable, as explained in the text. Credit: Dan Chiras.

FIGURE 5.10. Label on Utility Disconnect. Utility disconnects like the one shown here must, according to the National Electric Code, be properly labeled to warn electricians that even though the switch is open, the terminals can be live. Notice the surge protector on the side of the disconnect and the hole in the switch handle through which the utility inserts its lock should they need to lock a system out of the utility network for repair. Utilities typically require installers to mount the AC disconnect close to the utility meter. Credit: Dan Chiras.

solar electric systems online—have dropped this requirement. That's because synchronous inverters are designed to shut down if the grid goes down. Grid-tied PV systems will not backfeed onto a dead grid. Period.

Grid-tied inverters are designed to shut down if the voltage or frequency of electricity in the grid falls outside of preprogrammed settings. If there is an under voltage or over voltage or under frequency or over frequency, the inverter senses something awry in the utility line and shuts down as a safety measure. All grid-tied inverters carry an Underwriter's Laboratory listing—UL 1741. This designation indicates that the inverter will automatically disconnect in case of a power fluctuation.

Inverters primarily shut down under two conditions: brownouts and blackouts. A brownout occurs when electrical demand is extremely high, for instance, on a hot summer day when everyone's running an air conditioner. In such instances, voltage and frequency in the line may drop. The inverter senses this perturbation, and shuts down.

Inverters also shut down when blackouts occur—that is, when there is a complete loss of electricity due to a downed line or a transformer that's been struck by lightning.

As will be discussed in Chapter 6, grid-compatible inverters constantly monitor line voltage and frequency (the frequency and voltage of electricity on the grid). When they detect a change in either, for instance, a drop in voltage due to a power outage, the inverter automatically shuts down—and stays off until power is restored. When the grid appears to be back to normal, the inverter reconnects, but only after a five-minute period, just to be sure.

As you can see, lockable utility disconnects are redundant because UL-listed inverters meet interconnection safety standards and can be relied upon to disconnect if the grid goes down. The lockable disconnect is a backup to the safety features of the inverter. The problem large utilities have discovered is they don't have the personnel to shut down every PV system in their service area during a power outage. There may be several hundred systems throughout a city, maybe even thousands of them.

Rapid Shut Down

Although AC and DC disconnects allow service personnel to work on inverters without fear of shock, fire fighters petitioned for changes in the National Electric Code to provide additional safety. This led to a significant change in the NEC 2014. What they argued, and rightly so, was that although a power outage or switching off the AC disconnect both disable an inverter, the DC side of the wiring remains live

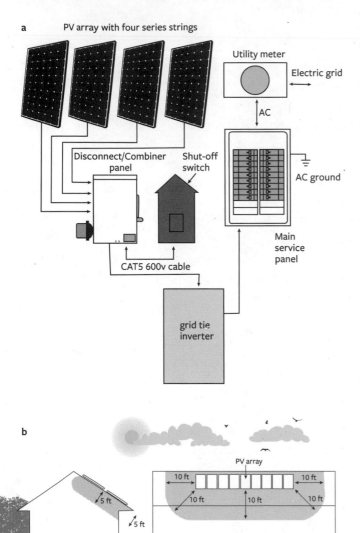

a PV array with four series strings

Utility meter

Electric grid

AC

Disconnect/Combiner panel

Shut-off switch

AC ground

CAT5 600v cable

Main service panel

grid tie inverter

b

PV array

10 ft   10 ft

10 ft   10 ft   10 ft

5 ft

5 ft

FIGURE 5.11. Rapid Shutdown Switch. (*a*) One of the newest requirements of the National Electric Code is the rapid shutdown switch. It is a safety measure lobbied by the firefighters of the United States to provide additional protection in case of a house fire. (*b*) This drawing shows the unprotected zones. Credit: Forrest Chiras.

as long as the Sun is shining. That is, the DC wires in a PV system will still have voltage. If severed by personnel fighting a fire, these wires could produce potentially lethal shocks. How would a wire become damaged by a fire fighters?

In some instances, fire fighters must cut through roofs to spray water on fires in attics. Were a fire fighter to cut the DC wires running from an array to a string inverter with an axe, he or she could be electrocuted. (Remember, the voltage in these wires is typically in the 300-to-400-volt range. Also bear in mind that the Code requires that any DC wires running into a building be encased in metal conduit, lessening the chance of damage and shock.)

Because of their concerns, the 2014 version of the National Electric Code required PV systems with string inverters to include a *rapid shutdown switch* or *RSS*. It enables fire fighters to terminate the flow of DC electricity within 10 feet (3.3 meters) of the array in installations where DC wires run across the surface of a roof or within 5 feet (1.6 meters) of DC wires entering a building, as illustrated in Figure 5.11a. By Code, the rapid shutdown switch must terminate the flow of electricity within ten seconds of activation.

As illustrated in Figure 5.11b, the rapid shutdown provision requires two components: a remote switch located near the service entrance and a DC disconnect. Rapid shutdown switches, can be manually or automatically engaged. Let's look first at the manually activated switch.

In a manually activated system, the shutdown switch (Figure 5.12a) connects to a DC disconnect at the designated location (Figure 5.12b). When a fire fighter flips the switch or pushes the button to shut down a PV system, his action opens the remote DC switch. (They are connected by a low-voltage

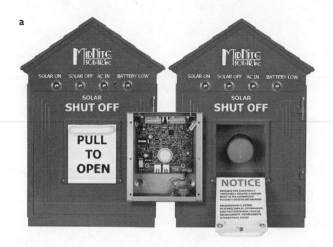

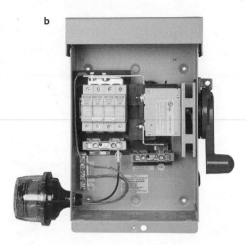

DC wire.) From the DC disconnect forward (toward the inverter), DC voltage is no longer present. Fire fighters can work without fear of shock or electrocution.

In contrast, automatic remote safety switches detect the drop in AC voltage that occurs when fire fighters pull a utility meter. It then sends a signal via a low-voltage DC wire to the DC disconnect, terminating the connection at the designated points.

RSS requirements can add a couple thousand dollars to a solar electric system installation (for equipment and labor) but are only required in systems with string inverters. They are not needed when installers employ microinverters. That's because microinverters are located at the array so there is no DC wire run from the array to the main panel. When the fire department pulls the electric meter, microinverters automatically shut down.

Net Metering and Billing in Grid-connected Systems

The idea of selling electricity to a local utility appeals to many people.

In most new installations, utility companies install digital net meters that tally electricity delivered to and supplied by a home or business. They keep separate totals of the electricity coming from and going to the grid.

All customers who connect their PV systems to the grid enter into a contractual agreement—called an *interconnection agreement*—with their utility. It spells out many details, including the provisions for paying a customer for surplus. This payment language is part of a net metering policy established by the state. Nearly every state has one. There are only a few holdouts, shown in Figure 5.13.

Net metering policies vary from state to state, but all provide rules regarding several key factors: (1) which types of systems the rules pertain to (for example,

FIGURE 5.12. Rapid Shutdown Switch and DC Disconnect. The 2014 National Electric Code requires (a) a rapid shutdown requires that is manually or automatically activated. (b) It operates a DC disconnect located within 10 feet of a solar electric array on a roof or 5 feet from the entry point of a DC circuit running from a PV array into a building. Credit: MidNite Solar.

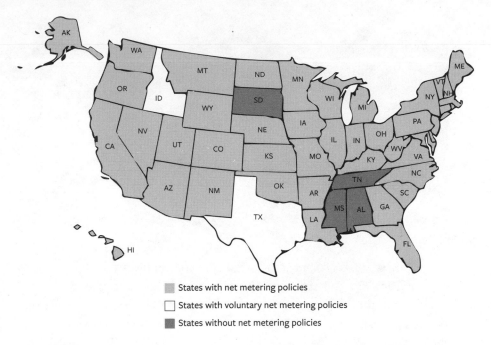

FIGURE 5.13. Map of Net Metering in the United States. This map shows which states currently have net metering policies. Credit: Forrest Chiras.

☐ States with net metering policies

☐ States with voluntary net metering policies

☐ States without net metering policies

wind-electric or solar-electric); (2) the maximum size a utility customer's system can be; (3) which utilities must abide by the rules (that is, municipal, investor-owned, or electric co-ops), and finally; (4) how customers are reimbursed for surpluses. Surplus electricity is referred to as *net excess generation (NEG)*.

Two types of net metering policies exist: monthly and annual. The "net" in net metering refers to the fact that the customer is billed only for net consumption, i.e., what remains after credits for surpluses are applied. In states like Colorado and Kansas that offer annual net metering, net excess generation is reconciled once a year. This billing arrangement is known as *annual net metering*.

Annual net metering is a lot like many cell phone plans. In a cell phone plan, surpluses can be carried from one month to the next for a year. If you have a surplus of 500 minutes one month, they can be used, free of charge, the next month.

In annual net metering, utilities carry surplus kilowatt-hours from one month to the next up to a year. Because of this, surpluses generated in summer months can make up for shortages in the fall. For example, if your system generates a surplus of 2,000 kilowatt-hours in the summer, and you need an extra 1,000 kilowatt-hours in November and another 1,000 kilowatt-hours in December, the electricity is yours for free.

So, what about surpluses, if any, that exist at the end of the annual billing period?

In annual net metering, unused electricity remaining in the account at the end of the year are handled in one of three ways. They are: (1) transferred to the utility (forfeited); (2) purchased at the retail price of electricity, i.e., the same price that the customer pays the utility for electricity; or (3) purchased by the utility at its wholesale rate. Wholesale rate is referred to as *avoided cost*. It what the electricity cost that utility—either to make it or buy it from another supplier. Avoided cost settlement is less desirable than reimbursement at retail. Once the account is reconciled, the balance is set to zero, and the net metering starts over for the following year. That said, some states (such as extremely renewable-energy-friendly Colorado) allow customers to carry their balance from year to year. Thus, if you have a surplus in 2017, you can carry it over to 2018. If you need it in 2019, it's yours for free.

The advantage of annual net metering is that it accommodates the seasonal variation in a PV system's production. In the summer months in many climates, PV systems produce more electricity than is consumed by a customer. In the winter months, PV systems typically produces less electricity than is consumed. Surpluses can be withdrawn from the "bank" and applied to bills in those months.

Consider an example: Suppose that a customer with a PV system mounted on her home delivered 800 kWh of electricity to the grid during the month of August, a fairly sunny month in most locations in North America. Suppose also that the customer consumed 600 kWh from the grid during that month. For this month, the customer would be credited with the net production of 200 kWh and would not be billed for any electricity, although she would be billed for the normal customer service charge and fees of $5 to $25. In this example, then, the customer would simply pay the customer service charge and applicable surcharges. She would carry the 200-kWh surplus over to September.

Let's assume that the next month, September, was unusually cloudy, so the customer delivered only 400 kWh of electricity, but the home consumed 550 kWh of electrical energy from the grid. In this month, the utility will deduct 150 kWh (the net consumption) from the customer's account, leaving a balance of 50 kWh. Since the balance in the account was sufficient to cover the net consumption, the customer would only be billed the monthly service charge.

Now consider the next month of net metering, October. Let's suppose in this month, the customer produced 600 kWh of electricity and consumed 800. The net consumption is 200 kWh, which exceeds the balance in the account. The utility would deduct 50 kWh from the account and bill the customer for the remaining 150 kWh.

In many states, utilities reconcile the customer's electric bills each month. This arrangement is known as *monthly net metering*. In monthly net metering, net excess generation is not carried from month to month. It is carried from day to day within a month. As a result, a surplus generated on Monday can be used on Friday. However, any surpluses remaining at the end of the month must be reconciled. They can't be carried over. To see how this works, consider the following example.

Suppose that the customer produced a surplus of 200 kWh of electricity in the first two weeks in the months of August. Because of heavy air conditioner use, however, the customer consumed 200 kWh more than the PV system produced during the second half of the month. In this case, the customer would not be charged for electricity. He or she would only pay the service fee.

Now suppose that when September rolls around, it's a particularly cold and sunny month, which is ideal weather for PV systems. The crystal clear blue days of September, in this example, result in a surplus of 500 kWh. That is to say, the customer produced 500 kWh of electricity more than she withdrew from the grid. Let's also suppose that the utility charges 15 cents per kilowatt-hour for electricity supplied by the grid and credits customers the same amount for surpluses delivered to the grid. If this system is in a state that offers monthly net metering and reimburses at retail rate (what customers pay), the solar customer would receive a check or credit for $75 for the surplus. (That's 500 kWh net excess generation, at 15 cents.)

If you're thinking that this could be profitable venture, don't get your hopes up. There aren't many states that reimburse customers at retail rates for monthly net excess generation. Most utilities pay for surpluses at the avoided cost—the cost of generating power. In Missouri, for instance, its monthly net metering policy allows utilities to reimburse at avoided cost. If you are paying the utility 10 to 12 cents per kilowatt-hour, they'll pay you about 2.5 cents. Some states like Arkansas simply "take" the surplus without payment to the customer—it all depends on state law. (If you're not happy with your state law, consider working to change it!)

Monthly net metering is generally the least desirable option, especially if surpluses are "donated" to the utility company or reimbursed at avoided cost. Annual reconciliation is a much better deal. It permits summertime surpluses to be "banked" to offset wintertime shortfalls. However, don't forget that even with annual net metering, the end-of-the-year surplus, if any, may be lost—forfeited to the utility, if state law permits this option.

The ideal arrangement, from a customer's standpoint, is a continual rollover. In such instances, there's no concern about losing your banked solar-powered kilowatt-hours. (I've found 13 states that permit continued rollover.)

Net metering is mandatory in many states, thanks to the hard work of renewable energy activists and forward-thinking legislators. At this writing (February 2016), 46 states and the District of Columbia have implemented statewide net metering programs. Only three states have no net metering policy at this writing: Mississippi, Tennessee, and South Dakota. Three utilities in these states have adopted net metering. Texas and Idaho have a voluntary program, which means the utilities can do whatever they want. Table 5.1 lists the states with the best policies. For a summary of net metering policy by state, see Appendix A. In Canada, some provinces have adopted net metering, too.

Although virtually all states have passed net metering policies, there are major differences. Differences include (1) who is eligible, (2) which utilities are required to participate, (3) the size and types of systems that qualify, and (4) reimbursement for NEG.

Eligibility is usually fairly broad. States allow homeowners and businesses of all sorts to participate. Most states also require net metering for all types of utilities.

Where policies vary the most is the size of PV systems businesses and homeowners can install. The size of systems varies widely, from 10 kW to 2,000 kW. Illinois, for instance, places a limit of 40 kW on PV systems while California allows systems up to 1,000 kW. New Mexico permits systems up to 80,000 kW.

Many states also limit size based on a customer's typical annual energy consumption. This gets a bit tricky. Colorado, for instance, allows residential customers to install up to 10 kW and businesses to install systems up to 25 kW, but the projected output of the system legally cannot exceed 120% of a customer's annual electrical demand. If 120% of a customer's demand could be met by a 6-kW system, that's the largest system he or she can install. There are ways around this, however. If, for example, you are getting married and your spouse is bringing two children to the marriage, you could make a case for a system larger than what you would need were you living alone.

Although they are reluctant to admit it, many utilities are finding that PV systems actually help them meet demand and do so quite economically. That's because residential and business PV systems often help "shave" peak load—that is, reduce daily electrical demand during periods of high use. As you can see in Figure 5.14, peak demand runs from about 11 AM to 6 PM. To meet electrical demand during this time, many utilities buy power from other utilities (purchase on the spot market) or fire up additional generating capacity, such as natural gas-powered electric plants. Both options can be very costly. Purchasing on the spot market (from other sources) can be very expensive. While a utility may generate electricity for 3 cents

Table 5.1. States with the Most Favorable Net Metering Policies

Arizona
California
Colorado
Connecticut
Delaware
Maryland
Massachusetts
New Hampshire
New Jersey
New York
Ohio
Oregon
Pennsylvania
Utah
Vermont
West Virginia

Source: "Best and Worst Practices in State Net Metering Policies and Interconnection Procedures," Freeingthegrid.org.

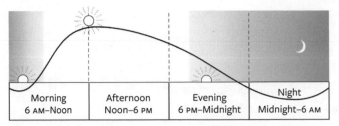

| Morning
6 AM–Noon | Afternoon
Noon–6 PM | Evening
6 PM–Midnight | Night
Midnight–6 AM |

FIGURE 5.14. Peak Demand. Peak load, that is, peak electrical demand, begins late in the morning and continues until the early evening in most regions of the United States. Solar electric systems help utilities shave peak demand, that is, reduce the demand for electricity during this time. This, in turn, makes it easier for utilities to supply inexpensive electricity to their customers.

Table 5.2. Pros and Cons of Batteryless Grid-tie Systems

Pros	Cons
Simpler than other systems	Vulnerable to grid failure unless a backup generator or an uninterruptible power supply is installed
Less expensive	
Less maintenance	
Unlimited supply of electricity (unless the grid is down)	
More efficient than battery-based systems	
Unlimited storage of surplus electricity (unless the grid is down)	
Greener than battery-based systems	

per kilowatt-hour, electricity purchased on the spot market (from other sources) can cost as much as 25 cents per kWh. Solar electricity is often generated in surplus while families are away at work or school, so PV systems can helps utilities reduce costs, saving the utility and customers money. Even if a utility reimburses PV customers for net excess generation at retail—say 12 cents/kWh—they are avoiding purchase on the spot market at 25 cents. It's a good deal for them.

The Pros and Cons of Grid-connected Systems

Grid-connected PV systems have their pluses and minuses, summarized in Table 5.2.

On the positive side, batteryless grid-connected systems are relatively easy to install. They require fewer components than other PV systems and are, therefore, considerably less expensive than other options. Batteryless grid-tie systems are often 30%–40% cheaper than off-grid systems and about 20%–30% cheaper than grid-connected systems with battery backup because batteryless grid-connected systems require fewer electrical components and no batteries. No batteries means no battery box or battery room, which can be costly to add to an existing home or to incorporate into a new home. Grid-tied systems also avoid costly battery replacement (something you can count on every seven to ten years, provided you take good care of your batteries). They also require much less maintenance than battery-based systems. In truth, maintenance is next to nothing. Although problems do crop up from time to time, they are extremely rare. (Be wary of installers who try to sell you a $300/year maintenance contract.)

Another advantage of batteryless grid-connected systems is that they can store months of excess production. Although they don't literally store excess electricity on site, like a battery-based system, they "store" surplus electricity on the grid in the form of a credit on your utility bill (in areas with annual net metering). The

Grid Storage: Is Electricity Stored on the Grid?

Many people are confused when renewable energy experts talk about storing electricity on the grid. There's good reason for this.

In a grid-connected system, surplus electricity is not physically or chemically stored on the grid. It is consumed by one's nearest neighbors.

Even so, electricity backfed onto the grid is "effectively" stored thanks to net metering. When surplus electricity is backfed onto the grid, a homeowner or business is credited for the surplus. The utility banks the credit. Although it sells the electricity to another customer, the utility keeps track of the amount of electricity a PV customer has delivered to the grid, so it can credit the customer later.

By keeping track of the surplus electricity produced by a net-metered renewable energy system, the utility says, "You've supplied us with x number of kWh of electricity. When you need electricity, for example, when the Sun is down, we'll supply you with an equal amount at no cost." In a sense, the utility has stored the electricity for its customer. Bear in mind, however, that when the utility gives back the electricity you've "stored" on the grid, they supply you with electricity likely generated by coal, nuclear energy, or natural gas. The net effect on your utility bill and the environment is the same, however.

grid also serves as an unlimited battery bank. Unlike a battery bank, you can never "fill up" the grid. It will accept as much electricity as you can feed it. In contrast, when batteries are full, they're full. They can't take on more electricity. As a result, surpluses generated in an off-grid PV system are typically lost. (In that situation, the PV array is temporarily disconnected from the battery bank so no current flows to the batteries when they are full.)

Another advantage of grid-connected systems is that the grid is always there—well, almost always. Even if the Sun doesn't shine for a week, you have a reliable supply of electricity. With an off-grid system, at some point the batteries run down. When they do, you need to run your backup generator—or sit in the dark. (For more on grid storage, see the sidebar "Grid Storage: Is Electricity Stored on the Grid?")

Yet another advantage of grid-connected systems with net metering is that utility customers suffer no losses when they store surplus electricity on the grid. With batteries, they do. As will be described in detail in Chapter 7, when electricity is stored in a battery, it is converted to chemical energy. When electricity is needed, the chemical energy is converted back to electrical energy. As much as 20% to 30% of the electrical energy fed into a battery bank is lost in converting from one form to another. In sharp contrast, electricity stored on the grid comes back in full. If you deliver 100 kWh of electricity, you can draw off 100 kWh. (This assumes annual net metering with the year-end reconciliation at retail price.) I'd be remiss if I didn't point out the grid has losses too, but net-metered customers get 100% return on their stored electricity.

Another advantage of batteryless grid-tie systems is that they are greener than battery-based systems. Although utilities aren't the greenest business in the world, they are arguably greener than battery-based systems. That's because batteries require an enormous amount of energy to produce. Lead must be mined and refined. Batteries must be assembled and shipped. Batteries contain highly toxic sulfuric acid and lead. Although old lead-acid batteries are recycled (there's at least one company in the United States [Deka] that recycles its own batteries), they're often recycled under less-than-ideal conditions. Many companies ship batteries to less developed countries, where children are often employed to remove the lead plates along the banks of rivers. Acid from batteries may contaminate surface waters.

When annually net metered at retail rates, batteryless grid-connected systems provide substantial economic benefits. As noted in Chapter 1, solar electricity has reached price parity with utility power in many parts of the world. That is, a solar system produces electricity for a customer at a cost equal to or less than the utility. In sunny sites, these systems may produce surpluses month after month. If the local utility pays for surpluses at the end of the month or end of the year, these surpluses can generate income that helps reduce the cost of PV systems and the annual cost of producing electricity.

On the downside of batteryless grid-connected PV systems is that they are vulnerable to grid failure. That is, if the grid goes down, so does your PV system. Even if the Sun is shining, a batteryless grid-tied PV system will shut down if the grid experiences one of several problem—for example, if an electric line breaks in an ice storm or lightning strikes a transformer at a substation, resulting in a power outage.

If power outages are a recurring problem in your area and you want to avoid service disruptions, you may want to consider installing a standby generator that automatically switches on if the grid goes down. Bear in mind, however, that a standby or backup generator takes many seconds to start and come online, so your power is interrupted during this time.

If you want to avoid this temporary interruption, you could install an uninterruptible power supply (UPS) on critical equipment, such as computers. A UPS consists of a battery pack and an inverter. If the utility power goes out, the UPS will supply uninterrupted power until its battery gets low. Or, as discussed in the next section, you may want to consider installing a grid-connected system with battery backup. In this case, batteries provide backup power during a power outage.

Yet another option is to install an inverter made by SMA, the Sunny Boy 5.0-US/6.0-US. These inverters are described by the company as a "secure power supply," although that's not entirely true. Here's what happens when the grid goes down: first, the inverter automatically terminates the flow of electricity to the main panel. This terminates the flow of electricity to all active loads and to the electrical grid. However, this inverter can continue to safely deliver up to 2,000 watts of power from a 5,000-watt to 6,000-watt PV array to a single electrical outlet wired directly into the inverter. Critical loads like refrigerators and freezers can be plugged into this outlet via an extension cord.

The only problem with this inverter is that it only supplies power during daylight hours. When the Sun goes down, the inverter shuts down and that outlet is out of commission until sunrise. I'll discuss this inverter in a little more detail in Chapter 6.

Grid-connected Systems with Battery Backup

Grid-connected systems with battery backup are also known as *battery-based utility-tied* systems or *battery-based grid-connected* systems—take your pick. They're all tongue twisters.

Grid-connected systems with battery backup ensure a continuous supply of electricity, even when an ice storm wipes out the electrical supply to you and huge swaths of your utility company's service area.

A grid-connected system with battery backup contains all of the components found in grid-connected systems: (1) a PV array, (2) an inverter, (3) AC and DC disconnects, including a rapid shutdown switch, (4) a main service panel, and (5) a utility meter or two to keep track of electricity delivered to and drawn from the grid. (Most of these components are shown in Figure 5.15.) Although grid-connected systems with battery backup are similar to batteryless grid-connected systems, they differ in several notable ways.

One of the most important differences is the type of inverter. Inverters in grid-connected systems with battery backup are a quite different from inverters in batteryless grid-tied systems. (I'll describe the differences in more detail in Chapter 6.)

As their name implies, battery-based grid-connected systems require a battery bank. Batteries for these systems are either flooded lead-acid batteries or, more

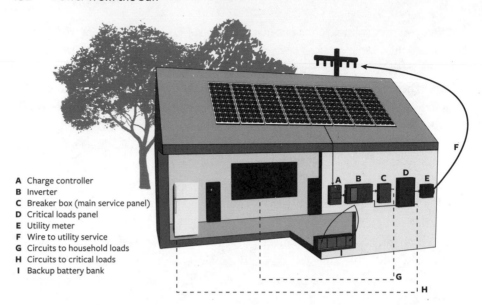

A Charge controller
B Inverter
C Breaker box (main service panel)
D Critical loads panel
E Utility meter
F Wire to utility service
G Circuits to household loads
H Circuits to critical loads
I Backup battery bank

commonly, low-maintenance sealed lead-acid batteries. Because batteries are discussed in Chapter 7, I'll highlight only a few important considerations here.

The first point worth noting is that battery banks in these systems are typically small—one-third to one-fourth the size of a battery bank required for an off-grid system. Small battery banks are employed because they are designed to provide sufficient storage to run just a few critical loads for a day or two while the utility company restores electrical service. Critical loads might include a few lights, the refrigerator, a well pump, the blower of a furnace, the pump in a gas- or oil-fired boiler, or a sump pump. Those who want or need full power during outages must install much larger and costlier battery banks or generators.

In battery-based grid-tie systems, batteries are called into duty only when the grid goes down. They're a backup source of power; they're not there to supply additional power, for example, to run loads that exceed the PV system's output. When demand exceeds supply, the grid makes up the difference, not the batteries. When the Sun is down, the grid, not the battery bank, becomes the power source.

It is also important to point out that battery banks in grid-connected systems are maintained at full charge—day in and day out—to ensure a ready supply of electricity should the grid go down. Keeping batteries fully charged is a high priority of these systems. Therefore, whenever power is being produced by the PV array, it first flows to the battery. (Batteries require a small amount of power to "float" at a fully charged state, described in Chapter 7.)

All of the power beyond what it takes to maintain this float charge is then sent to active loads in the home. When production exceeds consumption, the remaining electricity flows out, onto the grid.

In years past, some manufacturers designed their inverters to maintain a float charge (a small flow of current) to the batteries using utility power at night or during periods when the system was not producing power. This is also known as *trickle charging*. It's designed to counteract natural self-discharge that occurs in a battery as it sits idle. (You've probably witnessed this phenomenon with a car battery that sat idle for a long period.) Batteries require a continuous electrical charge and therefore become a regular load on a renewable energy system.

Today, manufacturers' inverters allow the batteries to rest at night and then resume trickle charging the battery pack to maintain full charge when the Sun comes up the next day.

Maintaining a fully charged battery bank requires a tiny amount of energy over long periods. That is to say, a small portion of the electricity a PV system generates is devoted to keeping batteries full at all times. This reduces system efficiency. Trickle charging could consume 5% to 10% of a system's output. In addition, trickle charging requirements increase as batteries age.

Battery banks in grid-connected systems don't require the careful monitoring they do in off-grid systems, but it is a very good idea to keep a close eye on them. When an ice storm knocks out your power, the last thing you want to discover is that your battery bank quietly died on you last year. For this reason, grid-connected systems with battery backup typically include a meter that monitors the total amount of electricity stored in the battery bank. Meters are usually built into charge controllers, discussed shortly. Pay attention to this meter. It will give you a reading of battery storage in amp-hours that indicates how full the battery bank is. I discuss this topic in detail in Chapter 7.

Another key component of grid-connected systems with battery backup is the *charge controller*, shown in Figures 5.16 and 5.17. Although I'll discuss charge controllers in depth in Chapter 7, a brief discussion is helpful at this point.

The charge controller performs several vital functions. As its name implies, a charge controller regulates the flow of electricity, ensuring batteries are rapidly and efficiently charged. Charge controllers also prevent batteries from overcharging. When a charge controller "sees" that the batteries are fully charged, it terminates the flow of electricity to them. As noted earlier, overcharging can permanently damage the lead plates in lead-acid batteries used in solar electric systems.

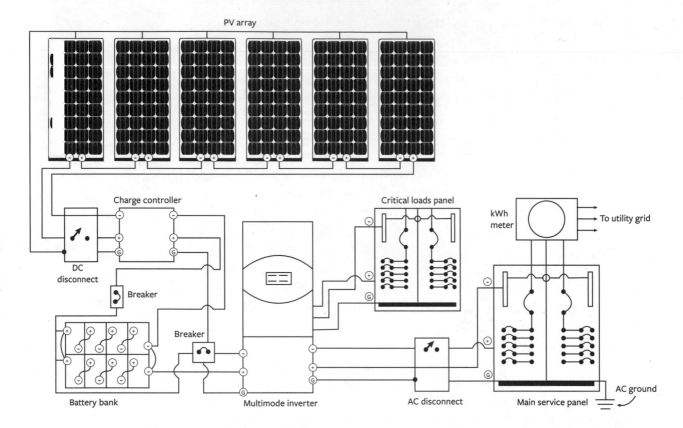

FIGURE 5.16. Schematic of Grid-connected System with Battery Backup. In these systems, the batteries are kept at full charge. Batteries are not called into duty unless there is a power outage. In that case, the system converts seamlessly to battery power. When this occurs, the inverter stops sending electricity to the main service panel, which prevents electricity from being backfed onto the grid. It only delivers electricity to the critical loads panel. Credit: Forrest Chiras.

Batteries also require protection from discharging too deeply, referred to as "over discharging." Over discharging can also damage the lead plates of batteries, dramatically reducing a battery's useful life. To prevent over discharging, charge controllers contain a low-voltage disconnect (LVD). It "covers" DC loads—that is, circuits that draw DC electricity directly from the battery bank to supply DC appliances and lights that are wired to the battery bank. To protect against over discharging by AC loads, PV systems rely on a low-voltage disconnect in the inverter. It shuts the inverter down if the battery voltage drops too low.

Low-voltage disconnects in the charge controllers and inverters terminate the flow of electricity out of the battery bank when the amount of electricity stored in batteries falls to 20% of the battery's storage capacity. In grid-tied systems with battery backup, deep discharge is rare. Over discharging typically only occurs during extended utility power outages—that is, when the utility power is down and the batteries are being used to supply critical loads for extended periods.

Modern charge controllers and inverters often contain a function called *maximum power point tracking* (MPPT). Discussed in detail in Chapter 7, maximum power point tracking circuitry optimizes the output of a PV array, thus ensuring the highest possible output at all times.

Modern charge controllers also contain a high-voltage/low-voltage DC conversion function. As you will learn in Chapter 7, this feature allows the array to be wired at much higher voltages than in previous years. In the 1990s, for instance, most battery systems were wired at either 12-, 24-, or 48-volts. If a system was wired at 24 volts, the installer had to wire the modules in 24-volt series strings, had to install a 24-volt charge controller, and had to install a battery bank wired at 24 volts. Even the inverter had to be designed to convert 24-volt DC electricity into 120-volt AC. Low voltages required short wire runs or very large and expensive copper wires to reduce line loss.

Today, new charge controller designs have radically changed the game. You can now purchase a charge controller, for instance, from MidNite Solar that will accept 150- or 200-volt DC electricity. Schneider Electric offers a charge controller that can accept up to 600-volt DC from a PV array. Higher voltages mean higher efficiencies, longer wire runs, and small-gauge wires, which are less expensive.

Modern high-voltage charge controllers convert high-voltage DC electricity from the array to a voltage that matches the voltage of the battery bank—typically 24 or 48 volts. A 600-volt charge controller, for example, would convert 600-volt DC from a PV array to 48-volt DC to charge a 48-volt battery bank.

Another key component of grid-tied systems with battery backup is the critical loads panel. As shown in Figures 5.15 and 5.16, all systems of this nature require a main panel as well as a subpanel, a separate breaker box that services critical loads. During installation in an existing home, branch circuits serving critical loads must be pulled from the main panel and placed in the critical loads panel.

Pros and Cons of Grid-connected Systems with Battery Backup

Grid-connected systems with battery backup enable homeowners and businesses to continue to operate critical loads during power outages. They may, for instance, allow homeowners to run their refrigerators and some lights while neighbors grope around in the dark and the food in their refrigerators begins to rot. These systems allow businesses to continue to use computers and other vital electronic equipment so they can continue operations while their competitors twiddle their thumbs and complain about financial losses.

Although battery backup may seem like a desirable feature, it does have some drawbacks. For one, grid-connected systems with battery backup cost more to install—about 30% more—than batteryless grid-tied systems. The higher cost, of course, is due to the installation of additional components, including the charge controller, critical loads panel, and batteries. Flooded lead-acid batteries and sealed batteries used in these and other renewable energy systems are expensive. Remember, too, that a battery bank needs a safe, comfortable home. If you are building a new home or office, you need to add a well-ventilated battery box or a special vented battery room that stays warm in the winter and cool in the summer to house your batteries. Battery banks generally need to be vented to the outside to prevent potentially dangerous hydrogen gas buildup, which can lead to explosions and fires if ignited by a spark or flame. Vented battery rooms and battery boxes add expense.

Because battery banks in these systems are typically fairly small, a homeowner or business owner will not be able to use them to meet all of their power requirements. Small battery banks may be taxed during extended power outages lasting more than a few days, especially if it's dark and cloudy.

Flooded lead-acid batteries also require periodic maintenance and replacement. As explained more fully in Chapter 7, to maintain batteries for long life, you'll need to monitor fluid levels regularly and fill batteries with distilled water every few months. Because battery banks in these systems are infrequently used, they tend to be forgotten. Out of sight, out of mind. If a homeowner fails to maintain batteries for several years, the electrolyte level can drop below the top of the lead plates, causing irreversible damage.

Even if well maintained, battery banks need to be replaced periodically. Typical batteries used in this system require replacement every five to ten years, at a cost of around two to three thousand dollars each time. (Batteries in grid-tied systems with battery banks tend to need replacement more often than batteries in off-grid systems.)

Yet another problem, noted earlier, is that battery-based grid-tie systems consume a portion of the daily renewable energy production just to keep the batteries topped off (fully charged).

Because grid-connected systems with battery backup are expensive and infrequently required, few people install them. When contemplating a battery-based grid-tied system, you need to ask yourself four questions: (1) How frequently does the grid fail in your area? (2) What critical loads are present and how important is it to keep them running? (3) How do you react when the grid fails? (4) Are there other less costly options, like a backup generator?

If the local grid is extremely reliable, you don't have medical support equipment to run, your computers aren't needed for business or financial transactions, and you don't mind using candles on the rare occasions when the grid goes down, why buy, maintain, and replace costly batteries?

In some cases, people are willing to pay for the reliability that a battery bank brings to a grid-connected system. One of my clients in British Columbia buys and sells stocks, bonds, and currency for huge accounts. He can't experience downtime during active trading—not for a second. As a result, he opted for a grid-connected system with battery backup for his home and office. See Table 5.3 for a quick summary of the pros and cons of battery-based grid-connected systems.

Table 5.3. Pros and Cons of Battery-based Grid-tie System

Pros	Cons
Provide a reliable source of electricity	More costly than batteryless grid-connected systems
Provide emergency power during a utility outage	Less efficient than batteryless grid-connected systems
	Less environmentally friendly than batteryless systems
	Require more maintenance than batteryless grid-connected systems

Off-grid (Stand-alone) Systems

Off-grid systems are designed for individuals and businesses that want to or must supply all of their needs via solar energy—or a combination of solar and wind or some other renewable source. As shown in Figure 5.17, off-grid systems bear a remarkable resemblance to a grid-connected system with battery backup. There are some noteworthy differences, however. Of course, there *is* no grid connection. As you can see in Figure 5.18, there are no power lines running from the house or business to the grid. These systems "stand alone." The main source of electrical energy in an off-grid system is the PV array. Electricity flows from the PV array to the charge controller. The charge controller monitors battery voltage and delivers DC electricity to the battery bank. When electricity is needed, it is drawn from the battery bank via the inverter. The inverter converts the DC electricity from the battery bank, typically 24 or 48 volts in a standard system, to higher-voltage AC, typically the 120 volts required by households and businesses. AC electricity then flows to active circuits in the house via the main service panel.

Off-grid systems often require a little "assistance" in the form of a wind turbine, microhydro turbine, or a gasoline or diesel generator, often referred to as a *gen-set*. One or more of these energy sources helps make up for shortfalls.

Although backup generators are commonly used in off-grid renewable energy systems, some experts, like wind-energy authority Mick Sagrillo, contend that properly sized PV/wind hybrid systems rarely, if ever, require them. In fact, he's retrofitted numerous PV systems with wind generators to avoid the need for generator backup. Even though hybrid systems work well, the majority of off-grid systems include generators. It takes a well-balanced mix of solar and wind resources to avoid the need for a gen-set. "A gen-set also provides redundancy," notes National Renewable Energy Laboratory's wind-energy expert Jim Green. Moreover, "if a critical

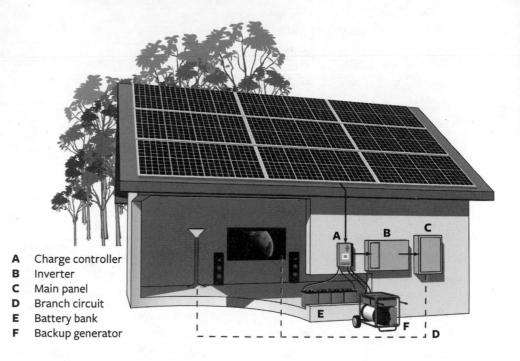

FIGURE 5.17. Off-grid System. Off-grid solar electric systems typically require a combiner box (not shown here), a charge controller, an inverter/charger, and a main service panel. A backup generator is often installed for battery maintenance and to provide power during unusually cloudy periods. Credit: Anil Rao.

A Charge controller
B Inverter
C Main panel
D Branch circuit
E Battery bank
F Backup generator

component of a hybrid system goes down temporarily, the gen-set can fill in while repairs are made." Finally, gen-sets also play a key role in maintaining batteries, a subject discussed in Chapter 7.

Off-grid systems with gen-sets also require battery chargers. (Not to be confused with the charge controller.) Battery chargers convert the AC electricity produced by the generator into DC electricity that's then fed into the battery bank.

In olden days, battery chargers were installed separately, but they're now built into inverters designed especially for these systems. (Inverters for battery-based systems are referred to as *inverter/chargers*.) When a homeowner fires up his generator, the inverter senses voltage at its input terminals; it transfers the home loads over to the generator through an internal, automatic transfer switch, and begins charging the battery from the generator.

Like grid-connected systems with battery backup, an off-grid system requires safety disconnects to permit safe servicing. A DC disconnect should be located between the PV array and the charge controller, and an AC disconnect should be installed between the inverter and the main service panel. A sign must be posted to warn fire fighters that a home contains an off-grid PV system with a battery bank.

Off-grid systems also require charge controllers to protect the batteries from overcharging with low-voltage disconnects (LVDs) to prevent deep discharge of the

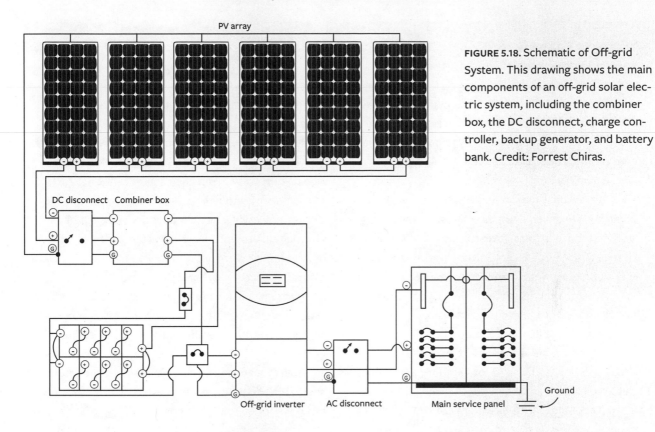

PV array

DC disconnect   Combiner box

Off-grid inverter   AC disconnect   Main service panel   Ground

battery bank. As a reminder, LVDs for DC loads are located in the charge controller, while LVDs for AC loads are housed in the inverter. Modern charge controllers also contain the maximum power point tracking circuitry mentioned earlier.

Charge controllers not only prevent batteries from being overcharged, they also prevent reverse current flow from the battery back to the array at night. Although reverse current flow is typically very small, it is best avoided. Modern-day charge controllers make this a non-issue.

As is evident by comparing schematics of the three types of systems, grid-tied systems with battery backup are the most complex. Second in terms of complexity is the off-grid system.

Off-grid systems can be partially wired for DC—that is, they contain DC circuits that are fed directly from the battery bank. DC circuits may be used to power lights or DC appliances such as refrigerators or DC well or cistern pumps. (For a discussion of DC circuits and DC appliances, see the sidebar, "DC Circuits in an Otherwise AC World?")

DC Circuits in an Otherwise AC World?

Most modern homes and businesses operate on alternating current electricity. However, off-grid homes supplied by wind or solar electricity—or a combination of the two—can be wired to operate partially or entirely on direct current electricity to power DC lights, refrigerators, televisions, and even ceiling fans. Why wire a home or cottage for DC?

One reason is that DC systems do not require inverters. Electricity flows directly out of the battery bank to service loads. This, in turn, can reduce the cost of the system, as household-sized inverters cost $1,000 to $4,000. However, cost-savings created by avoiding an inverter may be offset by higher costs elsewhere. For example, DC appliances typically cost more than AC appliances—considerably more. DC ceiling fans, for instance, cost four times more than comparable AC models. You could pay $300–$350 for a DC model, but $50–$150 for a comparable AC ceiling fan.

DC appliances and electronics are not only more expensive, they are more difficult to find. You won't find them at national or local appliance and electronics retailers.

Many DC appliances are tiny, too. Most DC refrigerators, for example, are miniscule compared to the AC models used in homes. That's because DC appliances are primarily marketed to boat and recreational vehicle enthusiasts, and there's not a lot of room in a boat or recreational vehicle for large appliances.

For these and other reasons, DC-only systems are rare. They are typically installed in remote cabins and cottages that are only occasionally used.

Another reason for avoiding the use of an inverter is efficiency. As you will see in Chapter 6, most inverters for off-grid systems are about 92% to 95% efficient.

That is, they consume some energy—5% to 8%—when converting DC to AC and boosting the voltage. Bypassing the inverter with a DC circuit to the water pump or refrigerator reduces this loss. Over the long haul, bypassing the inverter can result in large savings.

Although DC circuits may seem like a good idea, *Home Power*'s Ian Woofenden argues in favor of caution when considering this approach. He notes that water pumping typically does not require a lot of electricity, unless, of course you are irrigating a lawn or large garden or watering livestock. If you are living off grid, that's not very likely. Ceiling fans also are not that significant in the big picture. However, as Ian says, it's easier to make a case for a DC refrigerator or DC lighting, both more substantial electrical loads in our homes.

Even though modern refrigerators are much more energy efficient than their predecessors, they still are major energy consumers in homes. In off-grid homes, refrigeration often accounts for a substantial amount of a family's daily electrical demand. Because refrigerators consume so much electricity, a DC unit like those offered by Sun Frost or SunDanzer, can save a substantial amount of energy over the long run (Figure 5.19). Those thinking about an off-grid system, like one of my clients who runs a rustic "hotel" in Tulum, Mexico, with no grid power nearby, may want to consider DC refrigerators and DC freezers like the SunDanzer chest freezer.

While efficient, DC refrigerators and freezers do not come with the features that many Americans expect, such as automatic defrost or ice makers or cold-water dispensers on the door. They also cost much more than standard or even high-efficiency AC units. Because of these reasons, I rarely recommend the inclusion of DC circuits for off-grid installations. They're only for a certain type of renewable energy user—people who are trying to wring every possible kilowatt-hour from a small system.

There are other reasons to think twice about DC circuits in an otherwise AC home. For one, low-voltage DC circuits must be wired with larger gauge (thicker) wire. Copper wire is expensive, and the larger the gauge, the more you will spend. Richard Perez, founder of *Home Power* magazine, notes that electrical connections should be soldered, which adds to the cost of installation. DC circuits also require special plugs and sockets that are not readily available and are considerably more expensive than their AC counterparts.

Perez also points out that DC appliances are not typically as reliable or well built as their AC counterparts. They are, he says, primarily designed for intermittent use in recreational vehicles. DC appliances also wear out more quickly and thus require more frequent replacement.

As a case in point, a DC blender costs about twice as much as an AC model. The blender, Perez jokes, has only two speeds—on and off. Moreover, it can only be ordered from specialty houses by mail or via the internet. The DC blender he bought died after fewer than eight months of use. As if that's not enough to dissuade you from DC appliances, Perez points out that many appliances have no DC counterparts.

Arguing in favor of AC installation, Perez notes in an article in *Home Power* magazine, "the main advantage of using AC appliances is standardization. The wiring is standard—inexpensive, conventional house wiring. The appliances are standard, and are available with a wide variety of features." Furthermore, "the appliances are designed for regular use, and most are reliable and well built."

DC wiring may also need to be installed in metal conduit when it is run through walls, floors, or attics. This adds even more to the cost of a system.

Why metal conduit? DC electricity arcs—that is, it can jump across a gap fairly easily. If a wire is severed, for example, by a nail driven into a wall, sparks

FIGURE 5.19. Sun Frost Refrigerator. This Sun Frost refrigerator/freezer is superefficient and highly reliable, although quite expensive. (It comes in AC and DC models.) I've used one in my home in Colorado for many years. It uses about 275 kWh a year. If you shop carefully, you can find a 20 cubic foot refrigerator that uses less than 400 kWh a year. Credit: Sun Frost.

"jumping" across the gap could start a fire. (Be sure to check with your local building department or a licensed electrician to determine conduit requirements for DC circuits.)

While the main disadvantage of using AC appliances in off-grid systems is the cost of the inverter and the energy lost due to inversion inefficiency, modern inverters are remarkably efficient. Their operating efficiency is similar to or less than the amount of energy lost in low-voltage DC wiring, especially if the wire runs are long or are not terminated properly. In the final analysis, then, DC may not save any energy at all by bypassing the inverter.

If you are thinking about installing an off-grid system, your best bet is an AC system. Even so, you may want to consider installing a few DC circuits. In my off-grid home in Colorado, which is powered by solar electricity, I used AC electricity almost entirely. However, I wired one DC circuit to power a DC pump that pumps water from my cistern to the house. I added a second DC circuit to power three DC ceiling fans (although, after I priced them I opted for AC fans instead). At the time, I thought that I ought to install a DC refrigerator, but if you shop carefully you can find AC refrigerators that are nearly as efficient as the off-grid DC refrigerators.

You might want to consider installing a DC circuit in the utility room—or wherever the inverter is located—just to power a DC compact fluorescent light bulb in case of emergency! That way, if the system goes down at night, and you need to find out why, you'll have some light. Be sure to have a spare DC light bulb on hand, too.

Pros and Cons of Off-grid Systems

Off-grid systems offer many benefits, including total emancipation from the electric utility (Table 5.4). They provide the high degree of energy independence that many people long for. You become your own utility. In addition, if designed and operated correctly, they'll provide energy for many years. Off-grid systems also provide freedom from power outages.

Off-grid systems do have some downsides. As you might suspect, they are the most expensive of the three systems. Large battery banks and backup generators add substantially to the cost—often 50% to 60% more. In addition, you'll need space to house your battery banks and a generator. I had to build an insulated, soundproof ventilated shed to house my backup generator in Evergreen, Colorado, to appease neighbors who complained about the noise. Battery and generator "housing" add cost. Batteries

Table 5.4. Pros and Cons of an Off-grid System

Pros	Cons
Provide a reliable source of electricity	Generally the most costly solar electric systems
Provide freedom from utility grid	Less efficient than batteryless grid-connected systems
Can be cheaper to install than grid-connected systems if located more than 0.2 miles from grid	Require more maintenance than batteryless gridconnected systems (you take on all of the utility operation and maintenance jobs and costs)

also require periodic maintenance and replacement every seven to ten years, depending on the quality of batteries you buy and how well you maintain them.

Although cost is usually a major downside, there are times when off-grid systems cost the same or less than grid-connected systems—for example, if a home or business is located more than a few tenths of a mile from the electric grid. Under such circumstances, it can cost more to run electric lines to a home than to install an off-grid system. (I discussed this in Chapter 4.)

When installing an off-grid system, remember that you become the local power company, and your independence comes at a cost to you. Also, although you may be "independent" from the utility, you will very likely need to buy a gen-set and fuel, both from large corporations. Gen-sets cost money to maintain and operate, and you may be dependent on your own ability to repair your power system when something fails. Independence also comes at a cost to the environment. Gen-sets produce air and noise pollution. Lead-acid batteries are far from environmentally benign. As noted earlier, although virtually all lead-acid batteries are recycled, battery production is responsible for considerable environmental degradation. Mining and refining the lead, as noted earlier, are environmentally destructive. Thanks to NAFTA and the global economy, lead production and battery recycling are being carried out in many poor countries with lax or nonexistent environmental policies. They are responsible for some of the most egregious pollution and health problems facing poorer nations across the globe, according to small-wind-energy expert Mick Sagrillo. So, think carefully before you decide to install an off-grid system.

Off-grid systems also require a huge commitment to energy efficiency and a change in lifestyle. Don't think you can install a hot tub or ground-source heat pump or an electric water heater. Resistive heating and heat pumps are out of the question. They use way too much energy for a typical off-grid system. You'll need to eliminate all phantom loads as well and be especially vigilant about leaving lights and electronic equipment running when not in use. Every appliance or electronic device you purchase will need to be as energy efficient as possible. In short, the average energy-wasteful lifestyle is not amenable to an off-grid system. For more on this topic, see the sidebar, "To Stand Alone or Not to Stand Alone?"

Hybrid Systems

As you've just seen, you have three basic options when it comes to solar electric systems. Each of these systems can be designed to include additional renewable energy sources. The result is known as a *hybrid renewable energy system* (Figure 5.20).

To Stand Alone or Not to Stand Alone?

Many people speak to me about going off grid by installing a PV system, even though their homes or businesses are currently grid-tied. While this may sound like a glorious way to live your life, the off-grid option comes at a price. When Mick Sagrillo consults with individuals who want to go off grid for philosophical reasons, he cautions them to consider the ramifications of this decision, especially if they are philosophically opposed to utilities and concerned about the environmental impacts of grid-generated electricity. He does not at all condone how utilities operate, but from a purely environmental perspective, he thinks it is extremely difficult to justify a battery bank and gen-set system over grid connection.

I discussed the downsides of batteries earlier, but haven't mentioned the generator. The electric grid delivers electricity generated by coal-fired power plants at about 30%–33% efficiency and nuclear power plants at 20% efficiency. A Honda gen-set charging a battery bank operates at about 5% efficiency, Sagrillo contends. The emissions from that coal-fired plant are regulated by the EPA—more or less. Although there are also emission regulations for small engines (manufactured since September 1, 1997), the emissions per kilowatt-hour of electricity are far greater from the backyard generator than from the coal-fired power plant. In fact, if everyone had a gen-set in their backyard to meet their electrical demands, we'd all suffocate. Visit Cali, or Beijing, or any number of developing countries where generators are used to produce electricity and you'll see, smell, and choke on the result.

If you are thinking about going off grid, the responsible approach might be to install a hybrid system in which solar electricity is supplemented by some other clean, renewable energy technology such as wind or microhydro.

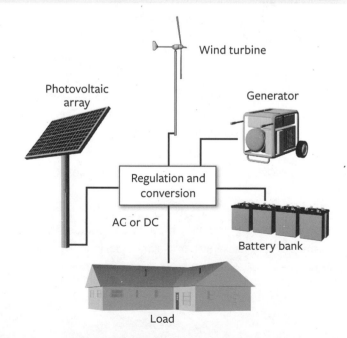

FIGURE 5.20. Hybrid PV and Wind System. Solar electric and wind energy systems can be installed along with a backup generator, if necessary. This is a simplified view of the system. Wind turbines and solar electric systems are generally wired to their own inverters, although there are a few inverters that can be wired directly to a hybrid solar electric and wind system. Credit: Anil Rao.

Hybrid renewable energy systems are extremely popular among homeowners in rural areas. For many years, one of the most active segments of the residential wind energy market consisted of PV system owners who incorporated wind energy, according to Mike Bergey of Bergey Windpower, which manufactures wind turbines.

Solar electricity and wind are a marriage made in heaven in many parts of the world. Why?

In most locations, solar energy and winds vary throughout the year. Solar radiation striking the Earth tends to be highest in the spring, summer, and early fall. Winds tend to be strongest in the late fall, winter, and early spring.

Table 5.5 shows that sunlight is relatively abundant in central Missouri from March through October. Table 5.6 shows that winds, however, pick up in October and blow through May. Together, solar electricity and wind can provide 100% of a family's or business's electrical energy needs. This complementary relationship is shown graphically in Figure 5.21.

In areas with a sufficient solar and wind resources, a properly sized hybrid PV/wind system can not only provide 100% of your electricity, it may eliminate the need for a backup generator. A wind system may also be used to equalize batteries in the winter. Equalization is a routine battery maintenance procedure that will help you greatly extend the lifespan of your batteries (discussed in Chapter 7). Because wind and PVs complement each another, you may be able to install a smaller solar electric array and a smaller wind generator than if either were the sole source of electricity. (Be sure to run the math very carefully.) You also need an excellent location and must install a turbine on a very tall tower—100 to 120 feet (30 to 36 meters) high! Do expect to pay a lot more to install a wind system to complement a PV system

Table 5.5. Solar Resource: Gerald, Missouri measured in kWh/m² per day

Lat 38.325 Lon –91.297	Jan	Feb	Mar	Apr	May	Jun	Jul	Aug	Sep	Oct	Nov	Dec	Annual Average
Tilt 38	3.66	3.86	4.71	5.43	5.28	5.50	5.75	5.66	5.48	4.79	3.27	2.95	4.7

Note: This data represents solar energy striking a collector mounted at an optimal angle for this location.

Table 5.6. Ten-year Monthly Average Wind Speed: Gerald, Missouri

Lat 38.325 Lon –91.297	Jan	Feb	Mar	Apr	May	Jun	Jul	Aug	Sep	Oct	Nov	Dec	Annual Average
m/s	5.96	5.93	6.50	6.31	5.27	4.83	4.40	4.31	4.64	5.16	5.67	5.40	5.40
m/hr	13.33	13.26	14.54	14.12	11.79	10.8	9.84	9.64	10.38	11.54	12.68	13.02	12.08

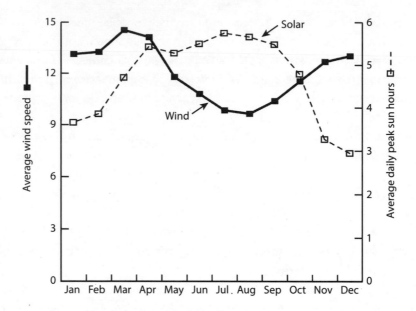

FIGURE 5.21. Complementary Nature of Wind and Solar. This graph of solar and wind resources at my educational center, The Evergreen Institute, and home in Gerald, Missouri, show how marvelously complementary wind and solar energy are. For off-grid systems, a combination of solar and wind can ensure a reliable amount of electricity year round. Be careful, however, as it may make more sense to simply expand a solar electric system.

than you would to install a small gas- or diesel-powered backup generator. Wind systems are not cheap!

If the combined solar and wind resource is not sufficient throughout the year or the system is undersized, a hybrid system will require a backup generator to supply electricity during periods of low wind and low sunshine. Gen-sets are also used to maintain batteries in peak condition and permit use of a smaller battery bank.

Choosing a PV System

To sum things up, you have basic choices when it comes to a PV system. If you have access to the electric grid, you can install a batteryless grid-connected system. It is by far the cheapest option. Or you can install a grid-connected system with battery backup. If you don't have access to the grid, you can install an off-grid system. All of these systems can combine two or more sources of electricity, creating hybrid systems.

When I consult with clients who are thinking of building passive solar/solar electric homes, I usually recommend a grid-connected system for those who live close to the utility grid. This configuration allows my clients to use the grid to store excess electricity, and it saves them a lot of money. Although they may encounter occasional power outages, in most locations in North America, these are rare and short-lived events. That said, grid-tied customers can install SMA inverters that provide a limited amount of daytime energy when the grid goes down. So long as the Sun is shining, these inverters can provide a couple thousand watts—sufficient to run critical loads.

If you are installing a PV system on an existing home or business that is already connected to the grid, it is generally best to stay connected. Use the grid as your battery bank. When installing such a system, however, be sure to contact the local utility company and apprise them of your intentions. You'll need to sign an interconnection agreement with the company.

Grid-connected systems with battery banks are suitable for those who want to stay connected to the grid, but also want to protect themselves from occasional blackouts. They'll cost more, but they provide peace of mind and real security. For best results, back up only the truly critical loads and make sure they are as efficient as possible. Some loads, like a forced air furnace, can be quite a challenge to back up. Not only is a furnace blower one of the larger loads in a home, it also runs during the time of the year with the least amount of sunlight.

Although more expensive than grid-connected systems, based on the cost per watt of installed capacity, off-grid systems are often the system of choice for customers in remote rural locations. When building a new home in a rural location, grid connection can be pricey—so pricey that an off-grid system makes good sense. As noted in Chapter 4, you will very likely be charged to run an electrical line to your home as well as for the meters and the utility disconnect. Expect to pay over $1,200 per electric pole (spaced every 300 to 400 feet) and even more for underground service. Hook up fees in such instances can run upward of $50,000 if you live a half-mile to a mile from the closest electric lines.

Before you decide to connect to the grid, be sure to check with your local utility. "Utility policies vary considerably when it comes to line extension costs," notes NREL's Jim Green. "Sometimes, the utility absorbs much of the cost in the rate base. Others pass most or all of the cost to the new customer."

CHAPTER 6

Understanding Inverters

The inverter is a modern marvel of electronic wizardry and an indispensable component of virtually all solar electric systems. Like batteries in off-grid systems, the inverter works long hours to provide us with electricity. As you know by now, its main function is to convert DC electricity generated by a PV array into AC electricity, the type used in our homes and businesses. In battery-based systems, inverters contain circuitry to perform a number of additional useful functions.

In this chapter, we'll examine the role inverters play in renewable energy systems and then take a peek inside this remarkable device to see how it operates. We'll discuss all three types of inverters and discuss the features you should look for when purchasing an inverter. But the first question to answer is this: Do you need an inverter?

Do You Need an Inverter?

Although this may seem like a ridiculous question, it's not. Some solar applications operate solely on DC power and, as a result, don't require inverters. Included in this category are small solar electric systems, like those in cabins, recreational vehicles, or sailboats—to power a few DC circuits. It also includes direct water-pumping systems that produce DC electricity to power a DC water pump. My DC solar pond aerator and the DC fans in my Chinese greenhouse are powered directly by PV modules.

All other renewable energy systems require an inverter. The type of inverter one needs depends on the type of system. Grid-connected PV electric systems require a utility-compatible inverter. Off-grid systems require inverters that are designed to operate with batteries. Grid-connected systems with battery backup require an inverter that's grid- *and* battery-compatible.

How Does an Inverter Create AC Electricity?

To help readers understand how inverters work, let's start with inverters installed in off-grid systems, known as *battery-based inverters*.

Battery-based inverters perform numerous functions. One of the most important is that they convert DC from the PV array to AC electricity, either 120 or 240 volts. In off-grid systems, electricity from the PV array is first fed into a charge controller. It then flows into the batteries. When electricity is needed, DC electricity stored in the batteries flows to the inverter. The inverter converts the DC electricity into AC electricity.

As noted in Chapter 5, batteries are wired so they receive, store, and supply 12-, 24-, or 48-volt electricity. The most common wiring configurations are 24 and 48 volts. In a 24-volt battery-based system, the inverter converts 24-volt DC from the battery bank to 120-volt or 240-volt AC electricity—either 50 or 60 Hz, depending on the region of the world you're in. (In North America, AC is 60 Hz; in Europe and Asia, it's 50 Hz.)

For many years, the DC-to-AC conversion and the increase in voltage were performed by two separate, but integrated, components in an inverter. The conversion of DC to AC occurs in an electronic circuit referred to as an *H-bridge*, for reasons that will be clear shortly. (It is also known as a *high-power oscillator* or *power bridge*). The second process, the increase in voltage, is performed by a transformer inside the inverter. As you shall soon see, many modern inverters have done away with the transformer. The new inverters are known as *transformerless inverters*.

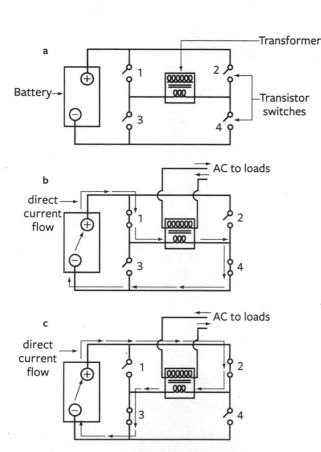

FIGURE 6.1. a (no electricity), b, and c: Diagram of Inner Workings of an Inverter. Credit: Anil Rao.

The Conversion of DC to AC

As shown in Figure 6.1a, the H-bridge consists of two vertical legs connected by a horizontal section, forming the letter H. Each leg contains two transistor switches that control the flow of electricity through the H-bridge.

To see how the H-bridge works, take a look at Figure 6.1b. As illustrated, DC electricity from the battery bank flows into the inverter. With switches 1 and 4 closed, low-voltage DC electricity flows from the battery into the vertical leg on the left side of the H-bridge then flows through the horizontal portion of the H-bridge and then down the lower part of the right leg of the H-bridge. It then returns to the battery.

An electronic timer then flips the switches, opening 1 and 4 and closing 2 and 3. As shown in Figure 6.1c, electricity now flows from the battery into the vertical leg of the H-bridge on the right side of the diagram through switch 2. It then flows through the horizontal portion of the H-bridge (in the opposite direction) and then down the lower part of the vertical leg on the left side through switch 3. From here, it flows back to the battery.

Opening and closing the switches very rapidly allows DC electricity to flow first in one direction and then the other through the horizontal portion of the H-bridge and transformer. This alternating flow is AC electricity. A small controller (not included in the figures) regulates the opening and closing of the switches in the H-bridge.

This clever device creates 12-, 24-, or 48-volt AC electricity. As you can see by studying the diagrams, however, it is contained within the inverter. But, how is the voltage increased? And how does the inverter transfer AC electricity to the main service panel?

Increasing the Voltage

As illustrated in Figure 6.1, the horizontal portion of the H-bridge forms loops (for simplicity, I've only included three loops). These loops are referred to as a *winding*. (A winding consists of single wire, wrapped many times around a core.) This winding lies next to a secondary winding. Together, the two windings form a transformer.

A transformer is a rather simple electrical device that can increase or decrease the voltage of AC electricity The transformer in a 24-volt inverter increases the 24-volt alternating current flowing through the H-bridge to 120-volt AC. The label "AC to loads" in Figure 6.1b and c indicates how the 120-volt AC electricity produced by the inverter exits. Those interested in learning how a transformer works should read the sidebar "How Transformers Work."

Square Wave, Modified Square Wave, and Sine Wave Electricity

An inverter, such as the one we've been examining, produces alternating current electricity. However, as shown in Figure 6.4a, it is very choppy. This graph is a plot of the voltage of the electricity over time. To understand what this graph represents, let's step back a second and review what we know about alternating current electricity. As noted in Chapter 5, electrons flow back and forth in a wire carrying

How Transformers Work

Transformers are used to increase (step up) and decrease (step down) voltage in the electrical grid and electrical devices. If you're tied to the electrical grid, chances are there's a transformer on a pole outside your home (Figure 6.2). This is known as a *step-down transformer* because it that decreases the voltage of the electricity in the electric line running by your home, which ranges from 7,500 to 24,000 volts (depending on the location) to 240 volts, the voltage used in homes. This transformer can also operate in reverse. That is, it can boost 240-volt AC backfed onto a grid from an inverter to match grid voltage. (That's one reason why inverters are designed to shut down when the grid goes down. This prevents high-voltage electricity from being backfed onto a dead grid.)

Step-up transformers, like the one found in an inverter, do just the opposite. That is, they increase the voltage.

Both step-up and step-down transformers use electromagnetic induction.

Electromagnetic induction is the production of an electric current in a wire as the wire moves through a magnetic field. This amazing phenomenon is the basis of virtually all electrical production technologies from coal- and natural-gas-fired power plants to wind turbines. For example, in the wind turbine shown in Figure 6.3, magnets are mounted on a rotating can, known as a *rotor*. It spins when the blades of the turbine spin. The magnets of a wind turbine rotate around a set of windings, numerous coils of copper wire. They are the stationary portion of a wind turbine generator, known as the *stator*. The rotation of magnets past the stationary (stator) windings produces electric current in the windings via electromagnetic induction. Cutting through the magnetic field literally pushes electrons in the copper wire, creating a current.

Magnetic fields are also produced as electricity flows through conductors. As you may recall from your reading, in an AC circuit, electrons flow back and forth, or cycle. As a result, in conductors carrying alternating current the magnetic field forms as the electrons race in one direction. When they stop, the magnetic field collapses. The electrons then reverse direction, and a new magnetic field forms. It collapses once again, though, when the electrons in the wire come to a stop at the end of the second half of the cycle. This alternating magnetic field—that is, expanding and collapsing magnetic field—induces an alternating current in a nearby wire, or in the case of a transformer, the secondary winding.

Transformers consist of two windings, one that's fed electricity, and another that produces electricity at a higher or lower voltage. AC electricity flowing through the first coil, known as the primary winding creates an oscillating (expanding and collapsing) magnetic field. This oscillating magnetic field induces an electrical current in the secondary winding. In other words, it has the same effect as moving a magnet back and forth past a wire. How does a transformer increase or decrease voltage?

The voltage in the secondary winding coil increases and decreases in proportion to the windings ratio, that is the number of turns in the secondary winding compared to the number in the primary winding. The greater the number of turns in the secondary winding, the higher the voltage. If the ratio of windings is 2:1, the voltage will double. If the ratio is 3:1, the voltage will triple.

If the ratio is reversed, that means there are fewer windings in the secondary coil than the primary coil, the voltage drops accordingly. If there are half as many windings in the secondary winding as the primary winding, the voltage drops by half.

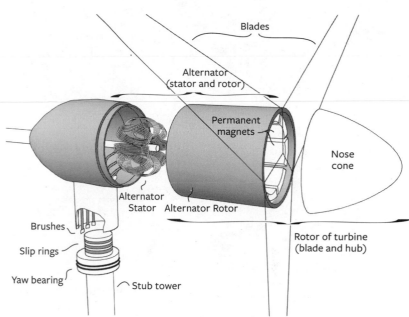

FIGURE 6.2. Step-down Transformer. This pole-mounted utility transformer reduces the voltage of incoming electricity, stepping it down from 7,500 to 22,000 volts (depending on the utility) to standard household voltage 240. This transformer steps up the voltage of electricity flowing from a PV array to the grid, as explained in the text. Credit: Dan Chiras.

FIGURE 6.3. Diagram of a Wind Turbine Showing Alternator. Small wind turbines, like the one shown here, contain a set of magnets that rotate the blades of the turbine in the wind. These magnets rotate around a several copper wire coils—the windings of the generator. As the magnetic fields pass by the windings, they generate AC electricity. Credit: Anil Rao.

alternating current at a rate of 60 cycles per second. If you measured the voltage in a wire carrying AC electricity with an instrument that responds fast enough, you'd see that the current changes over time. It has to because the flow of electrons shifts direction—albeit very rapidly. When the electrons are flowing one way, say to the right, the voltage rapidly climbs from 0 volts to +120 volts. The electrons momentarily stop, and the voltage then drops back to 0. However, the flow of electrons quickly shift direction, the voltage rapidly rises again to –120 volts. The graphs of voltage in Figure 6.4 trace the change in voltage over time. However, you will note that the voltage goes from 0 to +120 volts, back to 0, then up to –120 volts. What do the positive and negative signs mean?

The positive and negative signs refer to the "polarity" of the voltage. More precisely, it is a human construct that says the current changes direction. The electricity is not charged differently, it just travels in different directions. In Figure 6.4a,

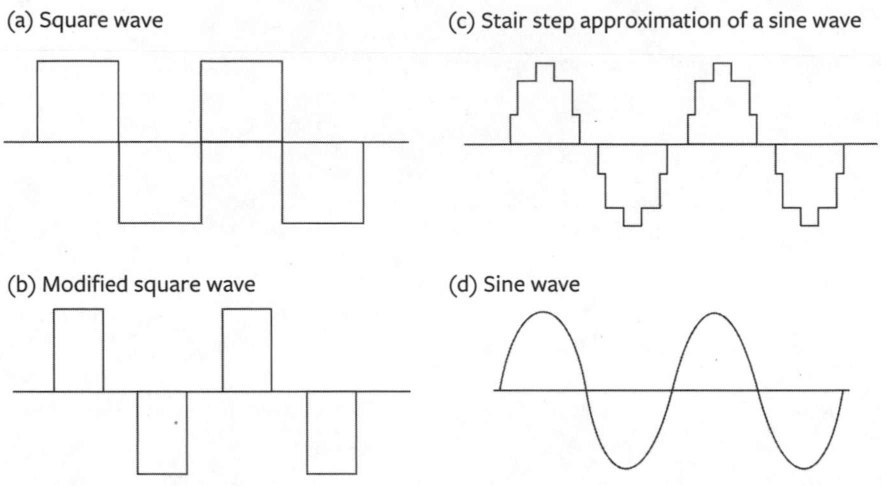

(a) Square wave

FIGURE 6.4. Electricity.

(c) Stair step approximation of a sine wave

(b) Modified square wave

(d) Sine wave

the +120-volt plateau means the electricity is flowing to the right and the voltage is 120. (The horizontal line is 0 volts; voltage readings above that line are considered positive, and readings below the line are negative.) The –120-volt simply signifies that the electrons are moving in the opposite direction for this brief moment.

The graph in Figure 6.4a indicates the voltage over an extremely brief period. The voltage shift from 0 to +120 volts then back down to 0 takes $\frac{1}{120}$ of a second; the voltage shift from 0 to –120 takes another $\frac{1}{120}$ of a second. Together, one complete cycle is $\frac{1}{60}$ of a second or 60 cycle per second electricity.

The basic H-bridge circuit found in some older inverters produces an almost perfect *square wave*. What that means is that the voltage jumps from 0 to +120 instantaneously, remains there for a short period, then instantly drops back to 0. Unfortunately, square wave electricity does not match the type of electricity produced by power plants. They produce *sine wave electricity*, shown in Figure 6.4d. Square wave electricity is not a very usable form of electricity. However, if this square wave is modified so that it pauses at zero volts briefly before reversing direction, the result is *modified square wave electricity*, shown in Figure 6.4b. Although doesn't appear to be much different, it is closer to a *sine wave*, the kind used in most homes. It can be used in off-grid homes, but has some serious shortcomings.

The problem with square wave and modified square wave electricity is that they contain *harmonics*. Harmonics are waveforms at frequencies that are multiples of the desired frequency. What that means is that there are electrons that cycle at different frequencies—besides the fundamental 60 cps.

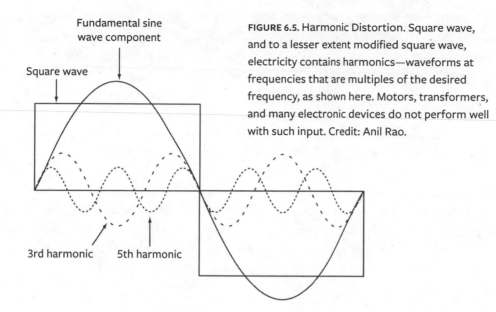

Fundamental sine
wave component

Square wave

3rd harmonic   5th harmonic

FIGURE 6.5. Harmonic Distortion. Square wave, and to a lesser extent modified square wave, electricity contains harmonics—waveforms at frequencies that are multiples of the desired frequency, as shown here. Motors, transformers, and many electronic devices do not perform well with such input. Credit: Anil Rao.

Square wave electricity contains odd harmonics—that is, frequencies that are 3-times, 5-times, 7-times, etc. the fundamental frequency of the square wave (Figure 6.5). The output of a 60 Hz square wave or modified square wave inverter, for example, contains not only 60 cps or 60 Hz, but also 180 Hz, 300 Hz, 420 Hz, etc. power. Motors, transformers and many electronic devices perform poorly in such circumstances. They can even be damaged by this phenomenon called *harmonic distortion* (see sidebar, "What Is Total Harmonic Distortion?").

Electronic filtering of the output of a square wave or modified square wave inverter can reduce these unwanted harmonics, but not eliminate them.

While modified square wave electricity works in many electrical devices like light bulbs, toasters, microwave ovens, and blenders, most devices run less efficiently on this type of electricity. Because of this, virtually all inverters sold in North America and Europe produce a much purer form of electricity known in the trade as *sine wave electricity*. As just noted, it is nearly identical to the type of electricity delivered via the utility grid. If you are planning on connecting to the grid, you must install a sine wave inverter. How do sine wave inverters produce such clean power?

Sine wave inverters contain multiple H-bridges, each of which produces square wave electricity (Figure 6.6). However, the H-bridges are stacked on top of one another, that is, carefully spaced to produce a stair-step waveform that closely resembles a sine wave, as shown in Figure 6.4c. What that means is that each H-bridge

What Is Total Harmonic Distortion?

Total harmonic distortion (THD) is a percentage of harmonics in an inverter output—the frequencies we don't want—in relation to the frequency we do want. Electricity from a square wave inverter has a THD of about 48%. The output of a modified square wave inverter is somewhat lower than this, although you won't find a value on the inverter specification sheet. (Inverter manufacturers don't like to talk about the THD of their modified square wave inverters.)

In sharp contrast, the THD of the output of a sine wave inverter is 5% or less, about the same as the distortion in utility power. In some cases, the THD of a sine wave inverter is lower than that in utility power.

The amount of distortion on power lines is dependent on where you live and what your neighbors—and the utility—are doing. Harmonic distortion is produced by variable speed electric motors in factories and electric welders, arc furnaces, and computers. They all produce short bursts of noise or distortion in the line. In homes and small businesses, harmonic distortion is produced by fluorescent lights, computers, and some power cubes. Harmonic distortion also arises from electric utility systems from overloaded transformers and other technologies. A perfect sine wave has 0% THD.

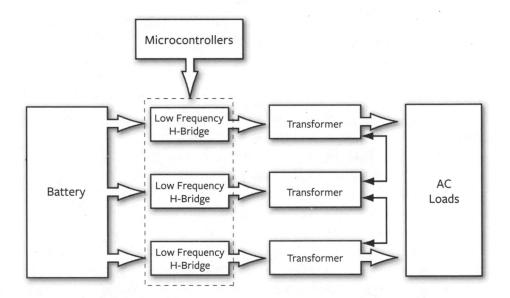

FIGURE 6.6. Multiple H-Bridges. Multiple H-bridges are found in transformer-based sine wave inverters. Each produces square wave electricity. Each H-bridge produces a different voltage AC electricity. The output of multiple H-bridges is combined to produce a stair-step approximation of sine wave electricity. It is filtered to produce pure sine wave electricity. Credit: Anil Rao.

creates a slightly different voltage. Their output is combined or stacked to form the stair-step waveform. Filtering this odd waveform produces a very clean sine wave, the kind of electricity our modern homes require (Figure 6.4d).

Transformerless Inverters

In recent years, many manufacturers have begun to transition to *transformerless inverters*. Unlike their predecessors, transformerless (TL) inverters rely on electronic components that convert DC to AC. Like so many things, the conversion is controlled by a computer inside the inverter. The process requires several steps.

> ## Modified Square Wave or Modified Sine Wave?
>
> A modified square wave inverter modifies—or cleans up—the square wave output from the H-bridge to reduce harmonics. Inverter manufacturers, however, usually label their modified square wave inverters as *modified sine wave inverters*. Why? It's marketing ploy. Customers know that a sine wave is better than a square wave, so an inverter that produces modified sine wave sounds better than one described as producing a modified square wave. In reality, they are the same.

Part of the impetus for this shift was to produce smaller, lighter inverters that cost less to ship and are easier to install than their heavier counterparts. (From my experience, though, they're still pretty heavy.) The other impetus was to reduce costs and increase efficiency. According to SMA, which produced its first TL inverter in 2010, because "transformerless inverters use electronic switching rather than mechanical switching, the amount of heat and humidity they produce is much less than standard inverters."

TL inverters are also capable of being wired into two different arrays with different orientations and/or tilt angles. Each array is serviced by its own power point tracking circuitry and software. It is as if the two arrays are separate PV systems. Traditional inverters only allow power point tracking of one array.

Transformerless inverters require slightly different wiring than standard inverters. For example, both the positive and negative wires from the PV array must run through a DC disconnect. In standard inverters, the positive wire is all that's required to be wired into a DC disconnect. In addition, both positive and negative wires must be protected by circuit breakers. In a standard inverter installation, only the positive wires must be protected by a circuit breaker.

Types of Inverters

Inverters come in three basic types: those designed for grid-connected systems, those designed solely for off-grid systems, and those made for grid-connected systems with battery backup. Table 6.1 lists alternative names for inverters and PV systems that you may encounter.

Table 6.1. PV System and Inverter Terminology

Common terminology used in this book (referring to systems and inverters)	UL 1741 (Refers to inverter)	NEC 690-2 (Referring to system)
Grid-connected	**Utility-interactive:** Operates in parallel with the utility grid.	**Interactive:** A solar photovoltaic system that operates in parallel with and may deliver power to an electrical production and distribution network.
Grid-connected with battery backup	**Multimode:** Operates as both or either stand-alone or utility interactive.	(Interactive, not separately defined)
Off-grid	**Stand-alone:** Operates independent of utility grid	**Stand-alone:** A solar photovoltaic system that supplies power independently of an electrical production and distribution network.
		Hybrid: A system composed of multiple power sources. These power sources may include photovoltaic, wind, microhydro generators, engine-driven generators, and others, but do not include electrical production and distribution network systems. Energy storage systems, such as batteries, do not constitute a power source for the purpose of this definition.

Grid-connected Inverters

Batteryless grid-connected PV systems require inverters that operate in sync with the utility grid. Two types are available: string inverters and microinverters. Both produce grid-compatible sine wave AC electricity. Both backfeed the surplus onto the grid. Moreover, the electricity they produce is indistinguishable from grid power; often, it's even a bit cleaner.

As you learned in Chapter 5, grid-connected inverters, also known as *utility-tie inverters*, convert DC electricity from the PV array into AC electricity (Figure 6.7). Electricity then flows from the inverter to the main service panel. It is then fed into active branch circuits, powering refrigerators, computers, stereo equipment, and the like. Surplus electricity, if any, is backfed onto the grid, running the electrical meter backward.

To produce sine wave electricity that matches the grid both in frequency and voltage, these inverters continuously monitor the voltage and frequency of the electricity on the utility lines. If an inverter detects a slight increase or decrease in voltage or frequency, it adjusts its output to conform. If, for example, the voltage drops from 243 volts to 242.7 volts or the frequency drops from 60 to 59.8 cps, the inverter automatically adjusts its output to conform to the grid. That way, electricity backfed from a PV system onto utility lines is identical to the electricity utilities are supplying to their customers.

Grid-compatible inverters are equipped with *anti-islanding protection*. Briefly mentioned in Chapter 5, though not named, anti-islanding consists of hardware and software that automatically disconnects the inverter from the grid in case of loss of grid power. That's its primary purpose. All grid-connected inverters are programmed to shut down if the grid goes down until service is restored. As noted in Chapter 4, grid-tied inverters terminate the flow of electricity to the electric lines to protect utility workers from electrical shock, even electrocution (fatal shock).

Grid-tie inverters also shut down if there's an increase or decrease in the frequency or voltage of grid power outside the inverter's acceptable limits. These conditions are referred to as *over/under voltage* or *over/under frequency*. If either the voltage or the frequency varies from the settings programmed into the inverter, it turns off. This can occur at the end of a utility line or near a manufacturing plant that has lousy power factor. (Power factor was defined in Chapter 4 and is explained in more detail

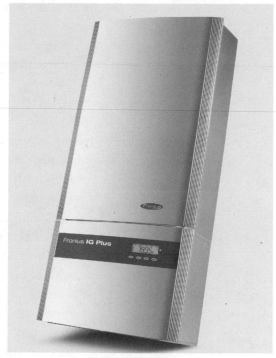

FIGURE 6.7. Fronius Utility-Intertie Inverter. This sleek, quiet inverter from Fronius is suitable for grid-connected systems without batteries. Fronius's inverters are produced at a new production facility in Sattledt, Austria, which started production in early 2007. The facility receives 75% of its electricity from a large, roof-mounted PV system and 80% of its heat from a biomass heating system. Credit: Fronius.

Is My Inverter Safe?

How does a buyer know if his or her inverter will disconnect from the utility during a power outage? It's something utility companies want to know. The answer is on the inverter nameplate. If it lists compliance with UL 1741 and IEEE 1547, you and your utility can rest assured that the inverter will automatically disconnect from the utility grid in case of an over/under voltage or over/under frequency. This feature is most important, of course, when the grid goes down completely (blackout). All grid-tied inverters today are UL 1741 certified. You or your installer will, however, have to inform the utility of this when you file the paperwork required to connect to the grid (the interconnection agreement).

there in the sidebar, "Real Power, Apparent Power, and Power Factor.") In such instances, grid power is poor enough that it doesn't match the window required by the inverter. The inverter remains off until the grid power returns within the acceptable range.

Grid-connected inverters also come with a *fault condition reset*—a sensor and a switch that turns the inverter on once the grid is back up or the inverter senses the proper voltage and/or frequency. Here's how it works: When an inverter detects an over/under voltage and/or frequency, it shuts off and remains off for 300 seconds (5 minutes). After 5 minutes, the inverter tests the voltage and frequency of grid power. If it has returned to normal, it will switch back on.

As noted in Chapter 5, one fact that many who are contemplating a grid-tied solar electric system often have difficulty accepting is that when the grid goes down or experiences over/under voltage or over/under frequency, not only does the inverter stop backfeeding surplus onto the grid, the entire PV system shuts down. As a result, the flow of electricity to the main panel and to active branch circuits in the house ceases as well.

As you know by now, the system shuts down to protect utility works from shock. It also shuts down because the inverter needs the grid connection to determine the frequency and voltage of the AC electricity it produces. Without the connection, the inverter can't operate. It's just wired that way.

To avoid losing power when the grid goes down, you can install a grid-connected system with battery backup. These systems allow homes and businesses to continue to operate when the utility grid has failed. Although inverters in such systems disconnect from the utility during outages, they continue to draw electricity from the battery bank to supply active circuits wired into a critical loads panel. It provides

How Does Power Factor Affect Voltage?

In this chapter you learned that voltage in an AC circuit, when graphed by a special electronic device, forms a perfect sine wave. If you graphed the amperage, you'd find the same thing. For optimum power, the voltage and current waveforms (frequency) should be aligned as closely as possible.

Power factor is a measure of how well the voltage and the current waveforms in electrical current are aligned. The greater their misalignment, the lower the power factor. (For the mathematically inclined, that's because $W = A \times V$.) A 95% power factor, typical of most utility power and inverters, means they are pretty close. In some cases, current waveform can lag behind the voltage waveform. For example, large electric motors cause the current waveform to lag behind the voltage waveform. When they are out of alignment, the real power (wattage) delivered to a load decreases.

electricity to the most important (critical) loads. This transition takes place so quickly and smoothly that you won't even know it occurred. In fact, the switch occurs so fast that even computers are not affected.

Another option, discussed in Chapter 5, is the SMA inverter that terminates flow to the main panel and the utility during power outages, but continues to produce electricity so long as the Sun shines.

Grid-connected inverters have LCD displays that provide information on a variety of parameters. The most important are the AC output of the inverter in watts and the number of kilowatt-hours produced by the system during a day. They also provide data on the total number of kilowatt-hours produced by a system since the inverter was installed.

Of lesser importance are the DC voltage and current (amperage) of incoming DC electricity and the voltage of the AC electricity backfeeding onto a grid. (One model, the Fronius inverter seen in Figure 6.7, keeps tab of the dollars saved and pounds of carbon dioxide emissions that you avoided by generating solar electricity.)

Another key feature found in many modern inverters—both string inverters and microinverters—is a function known as *maximum power point tracking* (MPPT) (first discussed in Chapter 3). MPPT is software and hardware that continuously tracks a solar array's DC voltage and amperage. It alters these parameters to create a combination of amps and volts that result in the maximum power output (in watts) of the array under all temperature and irradiance conditions. This feature ensures that the solar system produces the most power (watts) while it's converting DC to AC. In battery-based systems, MPPT is housed in charge controllers. The accompanying sidebar, "Maximum Power Point Tracking" explains how maximum power point tracking works.

Grid-connected inverters operate with a fairly wide DC input range. Inverters from Fronius and Oregon-based PV Powered Design are designed to operate within a DC voltage input range of 150 to 500 volts DC. This permits them to use a wider range of modules and system configurations. As noted in previous chapters, higher-voltage wiring allows us to transmit electricity more efficiently. It also permits the use of slightly smaller and less expensive copper wires. With the cost of copper skyrocketing, savings incurred by using smaller wire size can be substantial.

Another advantage of high-voltage wiring of arrays is that it allows installers to place arrays farther from the inverter than low-voltage arrays. My company has installed PV arrays from 100 to 1,000 feet from homes. (Really long runs like our 1,000-footers, however, require much larger wires to reduce line loss. We install either 100 or 200 amp aluminum wire for exceptionally long runs.)

Maximum Power Point Tracking

To get the most out of a PV array, modern solar electric systems incorporate a function known as maximum power point tracking (MPPT). The circuitry that allows MPPT is located in the inverter in grid-connected systems and in the charge controller of battery-based systems.

As noted in Chapter 3, the voltage and current produced by a PV array vary with changes in cell temperature, irradiance, and load. (You'll recall that a load is any device, equipment, or appliance that consumes electricity, inverters included.)

The relationship between current and voltage produced by a PV device (a cell, module, or array) is shown in Figure 6.8. This graph is known as a current-voltage curve or I-V curve (I represents amps and V represents volts). Take a moment to study the graph.

I-V curves show the relationship between current and voltage produced by a PV array—under a specific set of conditions, that is, a specific irradiance level and cell temperature. The PV array can operate anywhere along this curve.

To begin, look at the point on the curve labeled "short-circuit current." This is where the array operates if its output is short-circuited, i.e., the array's terminals are connected together. As indicated on the graph, the current is high, but the voltage is zero. *No power is being produced at this point.*

That's because power (watts) is determined by multiplying the voltage (volts) by the current (amps). Since the volts are zero at this point, the power (watts) output is zero.

Now look at the point labeled "open-circuit voltage." This is the point at which the array operates if its output is open circuited, i.e., not connected to anything. The current is zero and the voltage is at a maximum. Once again, at this point, the power output (watts) is zero.

At all points in between the short-circuit and open-circuit points, power is produced. How much power an array produces depends on where the array is operating on the curve. As the operating point moves away from the short-circuit point, the power output rises. The power reaches its peak near the middle of the "knee," the area where the curve changes from flat to steep. The point on the curve where the power reaches its maximum is called the maximum power point. As the operating point moves down the "knee," away from the maximum power point and toward the open-circuit point, the power decreases. What determines where an array operates on the I-V curve?

The answer is the load.

Just as the current and voltage of the PV array can be plotted on an I-V curve, so can the current and voltage of the load. Figure 6.9 shows the I-V curve of a typical load. As shown here, it is generally a straight line. As just noted, a load can operate anywhere on this line—at any combination of volts and amps. So, what determines where on an I-V curve a load will operate?

The answer is the array.

When the array and the load are connected, they

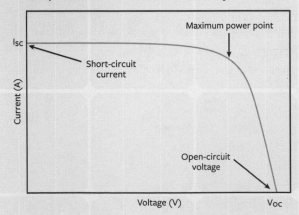

FIGURE 6.8. I-V Curve.

must operate at the same point—the one point that is common to both lines. This is illustrated by superimposing the I-V curves of the array and the load, shown in Figure 6.10. Where the two lines cross determines the operating point of the array when connected to the load, in our case, the inverter.

In order to get the maximum power out of the array, it must operate at its maximum power point. But, as we have just seen, the array will operate at a point determined by the load. So, how does the inverter get the load line to intersect the array I-V curve at the maximum power point?

Before we answer this question, I need to point out that the array I-V curve is not static. It changes with cell temperature and irradiance. Figure 6.11 shows how the array I-V curve changes with cell temperature. Take a moment to study this graph. It's very important that you do. Notice that as the temperature decreases, the voltage of the array increases. PV modules perform best when cold. Notice also that the maximum power point of the array shifts to the right.

Figure 6.12 shows how the array I-V curve changes with irradiance. Take a moment to study this graph as well. As you can see, as irradiance increases, the

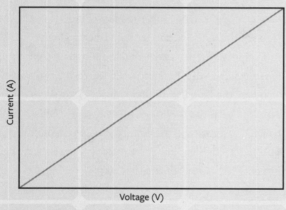

FIGURE 6.9. I-V Curve for Load.

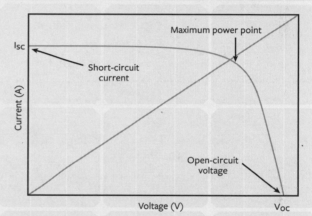

FIGURE 6.10. I-V Curve for PV Array and Load.

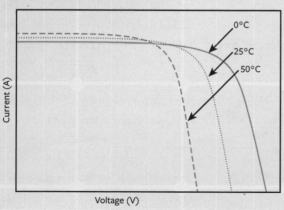

FIGURE 6.11. I-V Curve with Changing Temperature.

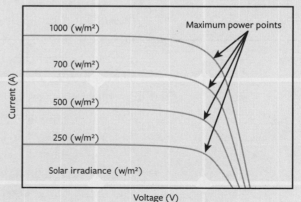

FIGURE 6.12. I-V Curve with Changing Irradiance.

current (amperage) of an array increases. That makes sense, because sunlight is the force that drives electrons out of a PV cell. The more intense the sunlight, the more electrons will be liberated. The more electrons liberated by a PV cell, the greater the current (amperage).

Temperature and irradiance vary during the day. On a cold winter day, for instance, irradiance is rather low first thing in the morning. The temperature is also low. During the day, irradiance and temperature both increase, changing the output of an array.

Because the voltage and current of an array change with irradiance and temperature, an array's maximum power point constantly shifts. It is a moving target. To get the load line to intersect the array curve at the maximum power point, the inverter must alter its load line to "track" the moving maximum power point of the array I-V curve. Fortunately, there is a way to do this.

The slope, or angle, of the load line can be altered by changing the impedance of the load. In general terms, impedance is a measure of how electricity flows through a material. More specifically, it is a measure of resistance to the flow of current in a conductor. A copper wire, for instance, poses greater resistance to electricity than an aluminum wire of the same size. Impedance is measured in ohms.

The impedance of the load is calculated by dividing the voltage by the current. In inverters, impedance (resistance to flow) in the wires coming into the inverter can be changed on demand via a circuit known as a *DC-to-DC converter*. This device changes the voltage of the DC electricity traveling from the array to the charge controller or inverter. As the voltage changes, the current also changes. As a result, the DC-to-DC converter can change the impedance and hence the slope of the load line so that it intersects at different points when irradiance and temperature shift. In these devices, MPPT controls the DC-to-DC converter, "aiming" the load line so that it always intersects the array curve at the maximum power point. The DC-to-DC converter and the circuitry that controls it is the MPPT. It's a brilliant bit of technology that can annually increase the output of an array by 15%.

Grid-connected inverters are installed indoors and outdoors. To help cool inverters installed in any location, they are equipped with screened openings that allow air to flow through to cool internal components. Inverters may also come with fans that automatically turn on when additional cooling is required. Inverters, like many other electronic devices, contain heat sinks. These fluted metal components draw heat off of devices and release it into the environment. In transformer-based inverters, the primary windings of the transformer and switches that control the flow of electricity through the H-bridge are connected to heat sinks to protect them.

Grid-connected inverters must comply with the utility specifications, including (1) acceptable total harmonic distortion, (2) over/under frequency detection, (3) over/under voltage detection, and (4) automatic shutdown in case of either. UL 1741 listing ensures that an inverter meets these and other utility requirements.

Numerous companies produce utility-intertie string inverters, that is, centralized inverters that service one or more strings of PV modules. These include Out-Back, Fronius, SMA, PV Powered Design, Schneider Electric (formerly Xantrex, which was formerly Trace), and SolarEdge. A number of companies also produce microinverters, module-scale inverters, including Enphase, Siemens (they make microinverters for Enphase), SMA, PowerOne, SolarEdge, and ABB.

Inverters have long been the weak point in grid-tied solar electric systems. Fortunately, manufacturers have made changes to enhance performance. Today, string inverters typically come with 5- or 10-year warranties. Microinverters, first introduced in 1993, though only more recently becoming popular, come with a 25-year warranty. To study inverter options in more detail, be sure to check out *Home Power* magazine. *Home Power* regularly publishes articles about the inverters available in North America.

Another product that you will encounter in your research is the *power optimizer*. These devices look a lot like microinverters and are wired directly to solar modules. Unlike microinverters, however, they do not convert module DC into AC. Rather, they contain MPPT technology that increases the output of each module before it delivers DC power to a specially designed string inverter. This maximizes the efficiency of the solar array. According to SolarEdge, optimizers last much longer than microinverters, although they come with the same 25-year warranty. Unfortunately, I have no experience with optimizers and don't know anyone who has installed them, so I cannot attest to their performance or compare them with microinverters.

As a final note, inverters should be installed inside buildings, preferably in dust-free, cool environments. While inverters are rated for outdoor installation and can tolerate cold as well as hot temperatures and moisture, most installers place their inverters outside for cost-savings, as shown in Figure 6.13a. I always install them in more hospitable environments to ensure a longer productive life. I can't help but think that exposure to extremes in weather as well as rain and snow will shorten an inverter's lifespan. As you shall soon see, inverters operate more efficiently when kept cool.

Off-grid Inverters

Off-grid solar systems and battery-based inverters were once the mainstay of the solar energy business. Over time, though, as solar electricity entered the mainstream, grid-connected systems have come to dominate the market. Nevertheless, there are a few companies that produce excellent off-grid inverters (Figure 6.14).

FIGURE 6.13. Inverter Installed Outside and Inside. Many installers place inverters that service large arrays near the array where they are exposed to the elements (*a*). I prefer to install inverters indoors like the one shown here (*b*). It provides added protection and ensures a longer lifespan. If you do install outdoors, be sure the inverter is shaded from the Sun throughout the year. The installer who put in the array shown on the left did shade the inverter, but only for six months of the year. Remember, the north side of an array will be illuminated early and late in the day from the spring equinox to the fall equinox. Be sure to protect against this sunlight, too. Summer sun will be the most intense. Credit: Dan Chiras.

FIGURE 6.14. Battery-based Inverter. This inverter manufactured by Schneider Electric is designed for use on the grid with batteries as well as off grid. For ease of wiring, you can purchase this or similar inverters from other manufacturers pre-wired. That is, they come with all the breakers and internal wiring on the DC and AC side. This makes installation a snap, and saves a lot of brain damage, as wiring these systems can be quite complex. Credit: Schneider Electric.

Like grid-connected inverters, off-grid inverters convert DC electricity into AC and boost voltage to 120 or 240 volts. Many off-grid homes are wired for 120-volt AC since it is impractical to install 240-volt loads like electric dryers, electric stoves, electric water heaters, electric space heat, or ground-source heat pumps. These devices consume way too much electricity for an off-grid system.

Off-grid inverters also perform a number of other functions as well, which we will explore in this section. If you're installing an off-grid system, be sure to read this carefully.

Battery Charging

Battery-based inverters used in off-grid and grid-connected systems with battery backup contain battery chargers. (Batteryless grid-tied inverters don't.) As their name implies, battery chargers charge batteries from an external source—either a gen-set in an off-grid system or the grid in a grid-connected system with battery backup. But isn't the battery charged by the PV array through the charge controller?

In off-grid and grid-tied systems with battery banks, batteries can be charged from the PV array or from an external source, either an AC generator or the electric grid. Battery charging from the PV array is regulated by the charge controller. It charges batteries with DC electricity. Battery chargers in inverters charge batteries from AC sources. They are a whole different animal. Battery charges in inverters convert AC to DC at the correct voltage. This is carried out by an electronic circuit containing devices called *rectifiers*. They're simple devices made from diodes that convert AC to DC.

In off-grid systems, battery charging gen-sets are used to restore battery charge after periods of deep discharge. This prolongs battery life and reduces damage to the lead plates. Battery chargers are also used during equalization, a battery maintenance operation required to increase the life of your batteries. You'll learn more about batteries and both of these processes in Chapter 7.

Battery charging from the electrical grid in a grid-tied system with battery backup occurs after the system has switched to battery power, for example, during a blackout. When the problem's been fixed and the grid is up and running again, the battery bank can be quickly charged using utility power via the battery charger in the inverter.

Abnormal Voltage Protection

High-quality battery-based inverters also contain programmable high- and low-voltage disconnect switches. These features protect various components of a system, such as the batteries, appliances, and electronics. They also protect the inverters.

The low-voltage disconnect (LVD) in an inverter monitors battery voltage at all times. When low battery voltage is detected (indicating that the batteries are deeply discharged) the inverter shuts off and often sounds an alarm. It no longer produces AC electricity. As a result, the flow of DC electricity from the batteries to the inverter stops. The inverter stays off until the batteries are recharged.

Low-voltage disconnect features are designed to protect batteries from very deep discharging, which can damage batteries. Although lead-acid batteries are designed to withstand deep discharges, discharging batteries beyond the 80% mark causes irreparable damage to the lead plates in a battery and leads to a battery's early demise. Although complete system shutdown can be a nuisance, the disconnect feature is vital to the long-term health of a battery bank.

Although the low-voltage disconnect feature is critical, *Home Power*'s Ian Woofenden notes that "it should not be used to manage one's batteries. It is designed more to protect an inverter from low voltage than to protect batteries from deep discharge. If the voltage of the battery bank reaches the disconnect level," he adds, "you may have already damaged your batteries." It's much better to develop an awareness of the state of charge of your batteries and learn to modify usage or turn on a generator well before the low-voltage disconnect kicks in.

To avoid the hassle of having to manually start a generator when batteries are deeply discharged, some inverters contain a sensor and switch that automatically activates a backup generator. When low battery voltage is detected, the inverter sends a signal to start the generator, provided the generator has a remote start capability. The fossil-fuel generator then recharges the batteries. AC electricity from the gen-set is converted to DC electricity by the inverter's battery charger. It then flows to the batteries, charging them.

When an AC generator is charging batteries, the inverter can send some of the AC electricity to the main service panel (in off-grid systems) or to a critical loads panel (in grid-connected sys-

A Cautionary Note

Home Power's Ian Woofenden notes that some renewable energy experts are leery about using auto-start generators. They prefer to have a human being in charge of the system, monitoring the charge of the batteries and the condition of the generator, including its fuel level. Why not automate? One reason is that if the inverter malfunctions, and it fails to shut down the gen-set when the batteries are fully charged, batteries can be overcharged and fried. In addition, generators not equipped with low-oil protection (a feature that turns the generator off if the oil levels are too low) can also be ruined if they're run under such conditions. Automatic is also not advised because it is more likely that a generator will run out of fuel, which can damage a generator.

tems) to supply active loads. As you might suspect, more sophisticated auto-start generators cost more than the standard pull-cord type.

High-voltage Protection

Inverters in battery-based systems also often contain a high-voltage shutoff feature. This sensor/switch terminates the flow of electricity from the gen-set when the battery voltage is extremely high. (Remember: high battery voltages indicate that the batteries are full.) High-voltage protection prevents overcharging, which can severely damage the lead plates in a battery. It also protects the inverter from excessive battery voltage.

Cost Factors

Inverters designed for off-grid systems cost a little more than the less complicated grid-tie inverters. However, the most significant savings in a batteryless grid-connected system comes from the omission of batteries and a battery room. A battery bank with twelve 6-volt batteries could cost upwards of $5,000—and that doesn't include wiring or construction of the ventilated battery box or battery room.

Another reason off-grid systems cost more is that batteries need replacement every seven to ten years, depending on the type of battery you install and how well you care for them. (If you manage your batteries carefully, you might be able to squeeze 10 to 15 years of service from them.) Deep cycling and allowing the batteries to sit partially discharged for long periods will shorten their lives. Avoiding batteries, therefore, reduces costly battery replacement. It also saves on the time required to maintain batteries, including equalization—a rejuvenating procedure I'll discuss in the next chapter. When comparing costs of an off-grid system to a grid-connected system, don't forget to factor in line extension and utility connection costs for the latter. An off-grid PV system may be much cheaper than a grid-tied system—if it avoids a $20,000 to $50,000 line extension fee.

Multimode Inverters

Grid-connected systems with battery backup are popular among homeowners who can't afford to be without electricity for a moment or those who experience frequent, long power outages. Grid-tied PV systems with battery backup require a special type of inverter: a battery- and grid-compatible sine wave inverter. They're commonly referred to as *multifunction* or *multimode inverters*.

Multifunction inverters contain features of grid-connected and off-grid inverters. Like a grid-connected inverter, they contain anti-islanding protection, which automatically disconnects the inverter from the grid in case of loss of grid power. They also contain over/under voltage and over/under frequency shutoff. In addition, they contain fault condition reset—to power up an inverter when a problem with the utility grid is rectified. And, like off-grid inverters, multimode inverters contain battery chargers and high and low-voltage disconnects.

Grid-connected systems with battery backup are the most difficult to understand. Many people think that the batteries are used to supply electricity at night or during periods of excess demand—that is, when household loads exceed the output of a PV array. That's not true. In a grid-tied system with battery backup, the grid is the source of electricity at night or during periods when demand exceeds the output of the PV array. The batteries are there in case the grid goes down. You don't want to use them each night. If the power goes out in the morning and your batteries have been used heavily at night, they'll be of very little use to you.

When batteries are full, multimode inverters act like any grid-tied inverter. They convert DC electricity from the array into AC that feeds active loads. Surpluses are backfed onto the grid.

If the battery voltage is low, for example, right after a power outage, the inverter charges the batteries and also powers active loads. It can charge the batteries and power AC loads from electricity drawn from the utility lines and, if the Sun is shining, electricity from the PV array. Once the battery bank is full, the system returns to normal operation.

If you are installing an off-grid system, you may want to consider installing a multifunction inverter in case you decide to connect to the grid in the future.

Grid-tied systems with battery backup are not the most efficient PV systems. That's because a portion of the electricity generated in such systems is used to keep the batteries topped off. (The batteries are trickle charged to offset self-discharge.) This may only require a few percent of a PV array's output, but over time, a few percent adds up. In systems with very old, poorly designed inverters or large backup battery banks, the electricity required to maintain the batteries can be substantial. The amount of energy required to trickle charge batteries also increases as batteries age. As batteries age, they become less efficient, so more power is consumed to maintain the float charge.

Retrofitting a Grid-tied System with Batteries

As noted in Chapter 5, many states have passed laws that allow utilities to charge solar electric powered customers more for electricity they purchase from the grid. This travesty has led many homeowners to contemplate going off grid—to cut their ties with the utility and become their own power company. How difficult is it to do this?

Going off grid can be challenging, because batteries store so little electricity and most people consume huge amounts of electricity on a daily basis—more than a typical solar battery bank can store. If you are one of them, switching to an off-grid system could be extremely difficult and will most certainly require a major life change.

If you are an energy miser and up to the challenge, you can replace your grid-tied inverter with an off-grid inverter and install a battery bank. If you installed microinverters, they'll need to be retired and you'll need to replace them with an off-grid inverter and battery bank. In either case, you'll also need to install a charge controller. But what if you just want to convert a batteryless grid-tied system to a grid-tied system with battery backup?

Homeowners have two options. One is called *DC coupling*, the other is *AC coupling*. Figure 6.15 illustrates the concept behind DC coupling. As shown in the drawing, DC coupling requires the replacement of your grid-tied inverter with a multimode inverter. It also requires installation of a charge controller, a battery bank, and a critical loads panel. In other words, it requires a complete overhaul. For details, I highly recommend Justine Sanchez's article entitled "AC-Coupling" in *Home Power* magazine, Issue 168, 2015.

AC coupling is an ingenious way of fooling a grid-tied inverter to continue operating when the grid goes down. It's a bit complicated, however. For details, see Figure 6.16. As illustrated, in this strategy you must install a battery bank and inverter-charger. The grid-tied inverter is then wired into a critical loads panel. The inverter-charger is wired into the main service panel that, of course, connects to the grid. So how does this work?

When the grid is up and running, the PV array produces DC electricity that is fed into the grid-tied inverter. The AC electricity it produces is

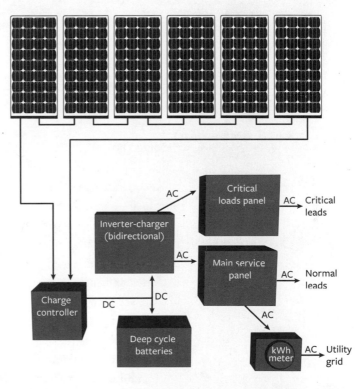

FIGURE 6.15. DC Coupling. This illustration shows how a grid-tied system can be modified into a grid-tied system with battery backup. This makeover requires a homeowner to scrap his or her grid-tied inverter and replace it with a brand new multimode inverter. Credit: Forrest Chiras.

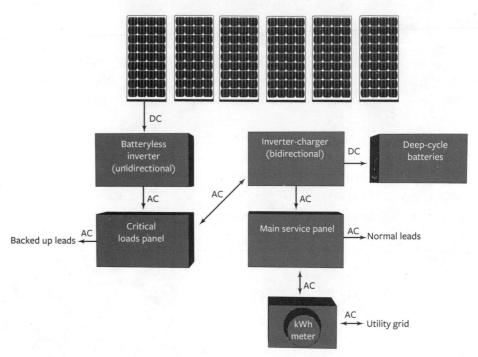

FIGURE 6.16. AC Coupling. This illustration shows another way a grid-tied system can be modified into a grid-tied system with battery backup. In this case, the homeowner keeps his or her batteryless grid-tied inverter, so long as it can be reprogrammed to work in this installation. He or she adds a multimode inverter, battery bank, and a critical loads panel. Credit: Forrest Chiras.

fed into the critical loads panel where it powers active loads. From this subpanel, unused AC electricity travels to the inverter-charger. From the inverter-charger, the AC electricity flows to the main panel where it supplies all the rest of active loads. Surpluses, if any, are backfed onto the grid.

If the power goes out, however, the inverter-charger severs its connection with the gird, as required by Code. It does so via an automatic shutoff switch (known as an *internal isolation relay*). The inverter-charger begins to draw electricity from the batteries. It feeds this electricity into the critical loads panel. The grid-tied inverter, in turn, senses normal voltage and continues to operate, converting DC power from the array into AC power. In essence, then, the inverter-charger tricks the grid-tied inverter into thinking that everything's okay. Because it and the inverter-charger only feed electricity into the critical loads panel, all is well.

Installing systems such as this can be tricky, so consult a professional who has experience in such matters. This configuration may void warranties on grid-tied inverters. Only specially programmed inverter-chargers can be used. For more information, go online and/or check out Justine Sanchez's article mentioned earlier. With these basics in mind, we turn our attention to the purchase of an inverter.

Buying an Inverter

Inverters come in many shapes, sizes, and prices. The smallest inverters, referred to as *pocket inverters*, range from 50 to 200 watts. They are ideal for supplying small loads such as VCRs, computers, radios, televisions, and the like. Most homes and small businesses, however, require inverters in the 2,500 to 12,000-watt range. Which inverter should you select?

Type of Inverter

If you are going to have a system installed, your installer will design it for you and specify the inverter and other components. Installers usually purchase solar equipment from a local or national distributor. They, in turn, typically carry inverters from one or two manufacturers. If the installer and distributor have been in business for a while, they carry an inverter that's reliable and backed by a good warranty.

Designing and installing your own system can be quite challenging. There is a lot to know, so this is a project best undertaken by those with extensive knowledge and experience in electrical wiring and those who have taken a few PV installation workshops.

The first consideration when shopping for an inverter is the type of system you are installing. If you are installing an off-grid system, you'll need an off-grid inverter/charger and a charge controller. If you are installing a grid-connected system with no batteries, you'll need an inverter designed for grid connection. The inverter must be matched to the output of the solar array. (Installers can help match the inverter to the array.) If you are installing a grid-connected system with battery backup, you must purchase an inverter that is both grid- and battery-compatible. Battery-based grid-tie inverters may also be a good choice for off-grid systems in case you decide to connect to the utility at a later date.

Even though a local supplier/installer may recommend an inverter, reading the following material will expand your understanding of inverters. It will also help you better manage and troubleshoot your PV system. If you are designing and installing your own system, you'll want to know as much as you can about inverters. Let's start with grid-tied inverters.

Choosing a Grid-tied Inverter

Shopping for a grid-tied inverter is relatively easy. If you know the size of your PV system, you simply need an inverter that matches it. If your system is rated at 5-kW, you need a 5-kW batteryless grid-tied inverter. This inverter will also service a 4.8 and 5.2-kW PV array.

After that, most people shop for efficiency. Modern grid-tied inverters tend to be around 96% efficient. Microinverters can operate at slightly higher efficiencies.

Be sure the inverter is UL 1741 listed. Also be sure to check out the warranty. Most inverters come with a 5-year warranty, although 10-year warranties are starting to become more common. In some instances, you may be able to purchase an extended warranty for a string inverter. Microinverters come with 25-year warranties.

It is important to check the output voltage of the inverter. Homes are typically wired at 240 volts. You need a 240-volt inverter. Commercial buildings are typically wired at 208, 480, or 600 volts. Be sure your inverter matches the voltage of your application.

When shopping for an inverter, be sure to study the DC voltage input range so you can calculate the length (number of modules) in each series string you can wire into that inverter. As you may recall, the number of modules affects the voltage of each series string.

Choosing a Battery-based Inverter

Choosing a battery-based inverter requires a little more know-how. In this section, I will discuss the key factors that go into choosing a battery-based inverter, including (1) input voltage, (2) waveform, (3) continuous output, (4) surge capacity, and (5) efficiency. I'll also discuss battery chargers, noise, and a few other factors to consider.

Inverter Input Voltage

When shopping for a battery-based inverter, you'll need to select one with a battery input voltage that corresponds with the battery voltage of your system. As noted earlier, battery banks in solar electric systems are wired either 12-, 24-, or 48-volts. Twelve-volt systems are common in cabins and summer cottages. Residential systems are either 24- or 48-volt systems, most commonly 48 volts, because they're the most efficient.

In the old days, PV systems had to be wired so that the voltage of all components matched. For example, a system with a 48-volt battery bank required an array wired at 48 volts, a charge controller wired the same, and an inverter wired to accept 48-volt DC (Figure 6.17). In systems such as this, 48-volt DC traveled from the PV array to the charge controller to the battery bank to the inverter. The inverter, in turn, converted the 48-volt DC into 120-volt or 240-volt AC.

Because all components of an off-grid renewable energy system had to operate at the same voltage, a 24-volt inverter would not work in a 48-volt system. Fortu-

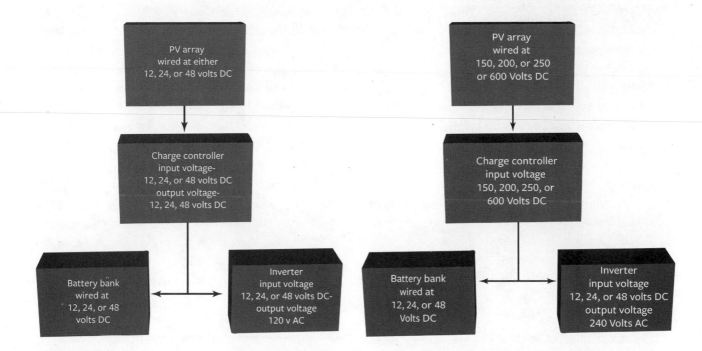

FIGURE 6.17. Old System Wiring Diagram. In the old days, off-grid systems had to be designed to so that all the components from the array to the inverter produced or operated with the same voltage. This translated into low-voltage wiring throughout. Because of this, these systems were not as efficient as they could have been, and installers had to place components as close as possible and use large-gauge wire to reduce line loss. Credit: Forrest Chiras.

FIGURE 6.18. New System Wiring Diagram. Today, off-grid systems can be designed differently—with high-voltage arrays. This is more efficient, allows placement of the array further from the batteries, and allows installers to use small-gauge, less expensive wiring in critical portions of the system. Credit: Forrest Chiras.

nately, things have changed. Several companies now produce charge controllers that can accept a much higher DC voltage from PV arrays. For example, Schneider Electric manufactures a 600-volt charge controller. PV arrays can now be wired up to 600 volts DC. This electricity is fed into the charge controller. It then converts the high-voltage DC to 48-volt DC that is fed into the batteries. The batteries feed the 48-volt inverter, that, in turn converts it to 120/240 AC (Figure 6.18). I'll talk more about this in Chapter 7, but for now, keep this in mind.

When shopping for an inverter for a grid-connected system, then, you'll need to find one that matches the battery bank voltage and the output of your charge controller. For more information, see the sidebar entitled "PV Array Voltage, String Size, and Choosing the Correct Inverter."

PV Array Voltage, String Size, and Choosing the Correct Inverter

In the 1970s through the mid-1990s, most off-grid PV systems were designed to charge 12-, 24-, or 48-volt battery banks. To meet the demands of this market, manufacturers produced modules rated at 12 volts, and installers built systems around this low-voltage module. In a 12-volt system, for instance, an installer would wire one or more 12-volt modules in parallel to produce the voltage needed to charge a 12-volt battery bank. (Wiring in parallel increases amperage but does not affect the voltage.) In 48-volt systems, installers would wire four 12-volt modules in series, then install one or more strings (a string is a group of modules wired in series) in parallel to produce higher amperage 48-volt DC electricity.

In these systems, the 12-, 24-, or 48-volt DC electricity was then fed to 12-, 24-, or 48-volt charge controllers. They delivered the electricity to appropriately sized battery banks. Appropriate inverters were installed to draw electricity from the 12-, 24-, or 48-volt battery banks. They converted the low-voltage DC electricity into 120/240-volt AC electricity suitable for household use.

Today, systems based on 12-volt increments have gone the way of the dinosaur. One reason for this is that most modern PV systems are batteryless grid-connected. Because there are no batteries, the inverters used in these systems can be designed to accept DC electricity from arrays at much higher voltages. In fact, most of grid-connected inverters operate with an input range of 150 to 550 volts, and with a maximum of 600 volts. (The National Electrical Code prohibits wiring PV arrays for homes higher than 600 volts.)

Higher voltage reduces losses as electricity flows from the array to the inverter. Because less current flows at higher voltage, higher-voltage systems permit the use of smaller gauge conductors, which reduces installation costs. Higher voltage also allows smaller components in inverters, reducing their size, weight, and cost.

To accommodate the higher-voltage inverters, manufacturers produce PV modules with higher nominal voltages—around 36 plus or minus a couple volts. These modules are then wired in series strings to produce high-voltage DC electricity for today's high-voltage inverters. To calculate the voltage of a string of modules, multiply the open-circuit voltage by the number of modules: $V_{oc} \times n$. If a module has an open-circuit voltage of 36.5 volts and you wire ten in series, the open-circuit voltage of the array is 365 volts.

To help installers determine how many modules they can wire in a string, virtually all inverter manufacturers provide online calculators. To calculate string size, simply enter the temperature conditions of the site (low temperature, described in the next paragraph) and the type of module you are going to install. The online calculator provides the maximum, minimum, and ideal number of modules in a series string (string length) for each of its inverters.

Temperature is important when determining the size of series strings in a PV array because, as you have learned in the previous chapter, the output of a PV array changes with temperature. Low temperatures, for example, increase the voltage of an array. If an array has not been sized carefully, the voltage of the incoming electricity could exceed the rated capacity of the inverter and the charge controllers—if it is a battery-based system. To determine the voltage of an array under coldest temperatures, designers use a midpoint between the average low and the all-time low for an area. From this, they can adjust the open-circuit voltage. In Missouri, for instance, the open-circuit voltage can be around 18% higher on really cold days. If the open-circuit voltage of the array just discussed is 365 volts (10 modules × 36.5 = 365), the highest voltage one would very likely ever experience would be 18% higher or 365 × 1.18 = 430.7 volts).

High voltage can seriously damage inverters, so be careful. This is most likely to occur early in the morning on cold, sunny days when the Sun first rises. At this time, the voltage will be at its peak.

High temperatures, in contrast, reduce the output of an array. If the voltage of a PV array falls below the rated input of a grid-tie inverter, a PV system will shut down and will remain off until the array cools down and the voltage increases. This, of course, reduces the efficiency of a system.

Many modern battery-based systems can be designed to handle higher-voltage arrays thanks to the introduction of maximum power point tracking (MPPT) charge controllers. MPPT charge controllers optimize power output of an array and accept higher-voltage input from PV arrays, but they contain step-down transformers that reduce the high-voltage array input to charge the battery bank at the appropriate voltage.

At first, these charge controllers were designed to accept 150 volts. They, in turn, could be used to charge 12-, 24-, or 48-volt battery banks. Newer charge controllers can be wired at 200, 250, and 600 volts DC.

A PV array's operating voltage is influenced by the ambient temperature and the number of modules wired in series, as just noted. Array temperature is also influenced by the type of mounting. As you'll learn in Chapter 8, some array mounts like ground-mounted racks allow more air to circulate around modules than others. Other mounting options like roof-mounted racks (standoff mounts) place PV modules in arrays very close to hot roof surfaces, which experience higher temperatures. The hotter the array, the greater the decline in its output. When calculating string size, then, an installer must also stipulate the type of mounting.

Designers must also take into account the slow degradation of modules, which reduces their output over time. Module output declines because of internal changes in the silicon. It may also decline as a result of moisture that seeps into the PV cells, which corrodes the internal electrical connections. It may also result from the slow deterioration of the ethylene vinyl acetate coating on the PV cells, discussed in Chapter 3. Because of this, it is important *not* to design a system that operates consistently near the lower end of the inverter's input voltage window. As the modules' output declines over time, you increase the chances of reduced energy output on hot days.

Waveform: Modified Square Wave vs. Sine Wave

The next inverter selection criterion you must consider is the output waveform. As noted earlier in this chapter, off-grid inverters are available in two types: modified square wave (often called modified sine wave) and sine wave. Grid-connected inverters are all sine wave so they match utility power. Multimode inverters for grid-tied systems with battery backup must be sine wave. (You can't backfeed onto an electrical grid unless you are sending sine wave electricity onto it.)

As you may recall, modified square wave electricity is a crude approximation of grid power but works well in many appliances such as refrigerators and washing machines and in power tools. It also works pretty well in most electrical devices, including TVs, lights, stereos, computers, and inkjet printers. Although all these devices can operate on this lower-quality waveform, they all run less efficiently, producing more heat and less work—light or water pumped, etc.—for a given input.

Problems arise, however, when modified square wave is fed into sensitive electronic circuitry such as microprocessor-controlled front-loading washing machines; appliances with digital clocks; chargers for various cordless tool; copiers; and laser printers. These devices require sine wave electricity. Without it, you're sunk. I, for example, found that my energy-efficient front-loading Frigidaire Gallery washing machine would not run on the modified square wave electricity produced by my very first inverter. The microprocessor that controls this washing machine—and other similar models (except the Staber)—simply can't operate on this inferior form of electricity. After I replaced the inverter with a sine wave inverter, I had no troubles whatsoever.

Certain laser printers may also perform poorly with modified square wave electricity. The same goes for some battery tool chargers, ceiling fans, and dimmer switches.

Making matters worse, some electronic equipment, such as TVs and stereos, give off an annoying high-pitched buzz or hum when operating on modified square wave electricity. Modified square wave electricity may also produce annoying lines on TV sets, and can even damage sensitive electronic equipment.

When operated on modified square wave electricity, microwave ovens cook more slowly. Equipment and appliances also run warmer and might last fewer years. Computers and other digital devices operate with more errors and crashes. The only time I have burned out a computer was when I was powering my off-grid home with a modified square wave inverter, and I've been working on computers since 1980. In addition to these problems, digital clocks don't maintain their settings as well when operating on modified square wave electricity. Moreover, motors don't always operate at their intended speeds. So why do manufacturers produce modi-fied square wave inverters?

The most important reason is cost. Modified square wave inverters are much cheaper than sine wave inverters. You will most likely pay 30% to 50% less for one.

Another reason for the continued production of modified square wave is that they are hardy beasts. They work hard for many years with very little, if any, main-tenance. (Their durability may be related to their simplicity: they are electronically less complex than sine wave inverters.)

Modified square wave inverters come in two varieties: high-frequency switch-ing and low-frequency switching units. High-frequency switching units are the cheaper of the two. A typical 2,000-watt high-frequency switching inverter, for example, costs 20% to 50% less than low-frequency models. They are also much lighter than low-frequency switching models, and are, therefore, easier to install. The high-frequency inverter may weigh 13 pounds compared to the 50 pounds of a low-frequency inverter. However, it is the latter that we typically use in household-sized PV systems.

Although low-frequency switching modified square wave inverters cost more and weigh more than their high-frequency switching cousins, they are well worth the investment. One reason for this is that they typically have a much higher surge capacity. That means they can deliver the greater surges of power that are needed to start certain electrical devices such as well pumps, power tools, dishwashers, washing machines, and refrigerators. More on surge capacity shortly.

Although there are some reliable modified square wave inverters on the mar-ket, I recommend that you purchase a sine wave battery-based inverter for off-grid systems. Their output is well suited for use in modern homes with their array of sensitive electronic equipment. SMA, Schneider Electric, and OutBack all produce excellent sine wave inverters.

Continuous Output

Continuous output is a measure of the power an inverter can produce on a con-tinuous basis—provided there's enough energy available in the system. The power output of an inverter is measured in watts, although some inverter spec sheets also list continuous output in amps (to convert watts to amps use the formula watts = amps × volts). Schneider Electric's (formerly Xantrex) sine wave inverter, model SW2524, for instance, produces 2,500 watts of continuous power. This inverter can power a microwave using 1,000 watts, an electric hair dryer using 1,200 watts, and several smaller loads simultaneously without a hitch. (By the way, the "25" in the model number indicates the unit's continuous power output; it stands for 2,500 watts. The "24" indicates that this model is designed for a 24-volt PV system.) The spec sheet on this inverter lists the continuous output as 21 amps.

OutBack's sine wave inverter VFX3524 produces 3,500 watts of continuous power and is designed for use in 24-volt systems. Off-grid homes can easily get by on a 3,000- to 4,000-watt inverter. Homes rarely require the full output.

To determine how much continuous output you'll need, add up the wattages of the common appliances you think will be operating at once. Be reasonable, though. Typically, only two or three large loads operate simultaneously. However, remember that, in some instances, multiple loads can operate simultaneously. A washer and well pump may be operating, for example, when a sump pump is running. If you are planning to operate a shop next to your home on the same inverter, you will need an inverter with a higher continuous output.

Surge Capacity

Electrical devices with motors, such as vacuum cleaners, refrigerators, washing machines, and power tools, require a surge of power to start up. To observe this, simply watch the amp meter in a renewable energy system when someone turns on a device such as a vacuum cleaner. Or watch the amperage on a Watts Up? or Kill A Watt meter, described in Chapter 4. If you pay close attention, you'll see a momentary spike in amperage. That's the power surge required to get the motor running. The spike typically lasts only a fraction of a second. Even so, if an inverter doesn't provide a sufficient amount of surge power, the power tool or appliance won't start! Not starting is not just inconvenient. Stalled motors draw excessive current and overheat very quickly. Unless they are protected with a thermal cutout, they may burn out.

When shopping for an inverter, be sure to check out surge capacities. All quality low-frequency inverters are designed to permit a large surge of power over a short period, usually about five seconds. Surge capacity or surge power is listed on spec sheets in either watts and/or amps.

Efficiency

Another factor to consider when shopping for inverters is their efficiency. Efficiency is calculated by dividing the energy coming out of an inverter by the energy going in.

While inverter efficiencies range from 80% to 95%, most models boast efficiencies in the mid to low 90s. A 94% efficient inverter, for instance, loses 6% of the energy it converts from DC to 120-volt AC.

It should be noted that the efficiency of inverters varies with load. Generally, an inverter achieves its highest efficiency once output reaches 20% to 30% of its rated capacity, according to Richard Perez, renewable energy expert and publisher of *Home Power* magazine. A 4,000-watt inverter, for instance, will be most efficient at outputs above 800 to 1,200 watts. At lower outputs, efficiency is dramatically reduced.

Cooling an Inverter

Because they're not 100% efficient, inverters produce waste heat internally. A 4,000-watt inverter running at rated output at 90% efficiency produces 400 watts of heat internally. (That's equivalent to the heat produced by four 100-watt incandescent light bulbs.) The inverter must get rid of this heat to avoid damage.

Inverters rely on cooling fins to passively rid themselves of excess heat to reduce internal temperature. Fins increase the surface area from which heat can escape. Some inverters also come with fans to provide active cooling. Fans blow air over internal components, stripping off heat. If the inverter is in a hot environment, however, it may be difficult for it to dissipate heat quickly enough to maintain a safe temperature.

Unbeknownst to many, inverters are also often equipped with circuitry that protects them from excessive temperature. That is, they're programmed to power down (produce less electricity) when their internal temperature rises above a certain point. As the internal temperature of the power oscillator (H-bridge) in a transformer-based inverter increases, the output current decreases. Translated, that means if you need a lot of power, and the internal temperature of your inverter is high, you won't get it. A Schneider Electric's SW series inverter, for example, produces 100% of its continuous power at 77°F (25°C), but drops to 60% at 117.5°F (47.5°C).

The implications of this are many. First, as I've pointed out before, inverters should be installed in relatively cool locations. If the inverter is mounted outside, make sure it is shaded. Second, inverters should be installed so that air can move freely around them. Don't box an inverter in to block noise (though not all inverters are noisy).

Battery Charger

Battery chargers are standard in battery-based inverters. As you may recall, a battery charger will allow you to charge your DC battery bank using AC from the utility (for grid-connected systems with battery backup) or an AC generator (for either off-grid systems or grid-connected systems with battery backup).

Noise and Other Considerations

Inverters can be installed inside a home or outside. While most solar inverters are rated NEMA-3R. NEMA stands for National Electrical Manufacturer's Association; 3R is a level of weather protection. It means they are contained in weather-tight enclosures that can withstand rain and snow and even ice forming on their surface. Even so, I believe that all inverters need to be protected from very low and very high temperatures. In fact, many grid-tie inverters are rated for installations with ambient air temperatures above −10°F.

Battery-based inverters are typically installed inside not so much for weather protection, although that's important, but so that they can be located close to the batteries. Wires from battery banks carry either 12, 24, or 48-volt DC. To reduce line loss, it is best to minimize line runs. Code requires the use of very large wires as well. (As noted, the batteries also need to "live" in a warm space.)

When installed inside, grid-tied inverters are almost always installed near the main panel. That's where the utility service enters a house. (Most inverter manufacturers recommend their equipment be housed at room temperature.)

If you are planning on installing an inverter inside your business or home, be sure to check out the noise it produces. Inquire about this upfront, or, better yet, ask to listen to the model you are considering in operation to be sure it's quiet. Don't take a manufacturer's word for it. My first inverter was described by the manufacturer as "quiet," but quiet compared to a jet engine. It emitted an annoyingly loud buzz heard throughout my house. The first six months after I moved into my off-grid, rammed earth tire home, the inverter's buzz drove me nuts, but I grew used to it.

Some folks are also concerned about the potential health effects of extremely low-frequency electromagnetic waves emitted by inverters as well as other electronic equipment and electrical wires. If you are concerned about this, install your inverter in a place away from people. Avoid locations in which people will be spending a lot of time—for example, don't install the inverter on the other side of a wall from your bedroom or office.

While we are creating a checklist of features to consider when purchasing an inverter, be sure to add ease of programming. My very first Trace inverter (DR2424, modified square wave) was a dream when it came to programming: all of the controls were manually operated dials. To change a setting, all I had to do was turn the dial. Its replacement, a Trace PS2524, which was a sine wave inverter, worked wonderfully but was challenging to program. Digital programming can be extremely complicated and the instructions can be difficult to follow.

My advice is to find out in advance how easy it is to change settings, and don't rely on the opinions of salespeople or renewable energy geeks that can recite pi to the 20th decimal place. Ask friends or dealers/installers for their opinions, but also ask them to show you. You may even want to spend some time with the manual (often available online) to see if it makes sense *before* you buy an inverter.

Another feature to look for in a battery-based inverter is power consumption under search mode. What's that?

The search mode is an operation that allows a battery-based inverter to go to sleep, that is, shut down almost entirely in the absence of active loads. The search mode saves energy (inverters may consume 10–30 watts or more from the battery when in the "on" mode).

Although the inverter is sleeping when it is in search mode, it's sleeping with one eye slightly open. That is, it's on the alert should someone switch on a light or an appliance. It's able to do this by sending out tiny pulses of electricity approximately every second. They are sent through the electrical wires of a home searching for an active load. When an appliance or light is turned on, the inverter senses the load and quickly snaps into action, powering up and feeding AC electricity.

The search mode is handy in houses in which phantom loads have been eliminated. (As you may recall, a phantom load is a device that continues to draw a small electrical current when off.) According to the Department of Energy, phantom loads, on average, account for about 5% of a home's annual electrical consumption. (In some homes, however, they can be as high as 10%.) Eliminating phantom loads saves a small amount of electrical energy in an ordinary home. In a home powered by renewable energy, however, it saves even more because supplying phantom loads 24 hours a day requires an active inverter. An inverter may consume about 30 watts when operating at low capacity. Servicing a 12-watt phantom load, therefore, requires an additional 30-watt investment in energy in the inverter—24 hours a day.

Energy savings created by eliminating phantom loads and continuous inverter operation does have its downsides. For example, automatic garage door openers may have to be turned off for the inverter to go to into the search mode. In my off-grid home in Colorado, I turned off my garage circuit at the breaker box or main service panel when I'd leave home. It was a pain in the neck, however, because when I'd arrived back home, I had to switch it back on.

Another problem occurs with electronic devices that require tiny amounts of electricity to operate, like cell phone chargers. When left plugged in, cell phone chargers may draw enough power to cause an inverter to turn on. Once the inverter starts, however, the device doesn't draw enough power to keep the inverter going.

As a result, the inverter switches on and off, *ad infinitum*. I found the same with my portable stereo.

Because of these problems, many people simply turn the search mode off so that the inverter keeps running 24 hours a day. Another option is to set the search mode sensitivity up, so it turns on at a higher wattage. However, most modern homes have at least one always-on load, for example, hard-wired smoke detectors or garage door openers that require the continuous operation of the inverter.

Computer Interfacing and Monitoring

When I wrote the first edition of this book, just a few years ago, only the more sophisticated—and most expensive—inverters came with remote monitoring. That is, they had electronics that kept track of energy production and could send that information to a computer or to a website that could be accessed by computer.

Today, virtually all grid-tied inverters come with a data logging capability. This data can be either uploaded to a computer or to a website. Most commonly, data is stored and displayed online at the manufacturer's website. However, you can subscribe to third-party monitoring services. (I've listed them in the Resource Guide at the end of the book.) As a result, a homeowner can check up on his or her solar system anywhere in the world that has internet access. For example, when you are vacationing in Mexico, you can go online to see how your system is performing. (Although you really should be swimming in the ocean or enjoying a margarita.)

Monitoring became popular with the introduction of Enphase microinverters. Their microinverters contain hardware (circuitry) and software that keep track of each module's output. This information is stored in the microinverter's memory but is also sent to a monitor known as an Envoy. The Envoy displays key information like the output of the array at that moment and lifetime production. It also passes info to a router then to a modem. From the modem, the information streams to the company's website via the internet, where it can be accessed by the installer and the homeowner.

Remote monitoring is a lot of fun but also quite useful as it allows homeowners and installers an opportunity to monitor systems for potential problems. I go online every month or so to check microinverter-based systems I've installed for customers. Online, you can see the wattage of each at any given moment and how much electricity each module has produced that day. The website even displays a graph of the daily output of a PV system.

Other monitoring systems are available, but they are all basically the same. They consist of data loggers and communications gateways. Even though I install remote monitoring on most of my systems, I always install a separate customer-owned utility meter as a backup. (Utility companies will often give you old utility meters that you can install yourself for your own use.) This meter is placed between the inverter and the main panel. It keeps track of the PV's cumulative output. That way, if anything goes wrong with the computer-based monitoring system, you will know what your system has produced. It's come in handy many times in my career.

Stackability

In the first edition of this book, I recommended that when buying a battery-based inverter a homeowner select one that can be stacked—that is, connected to a second or third inverter of the same kind in parallel or series. That's because the original battery-based inverters were small. They only produced 120-volt AC electricity. If you wanted 240-volt AC, you had to wire their AC output in series. Figure 6.19 shows four 120-volt OutBack inverters wired in a combination of series and parallel.

Although 120-volt inverters are still on the market, manufacturers now produce impressive inverters that produce either 120 or 240. There's no need to stack them.

Stackability and the 120/240-volt inverters allow one to produce 240-volt electricity to operate appliances such as electric clothes dryers, electric stoves, central air conditioning, or electric resistance heat. As a general rule, I recommend that you avoid such appliances, especially if you're considering an off-grid system. That's not because a PV system can't be designed to power these loads, but rather because they tend to use lots of electricity and you'll need to install a much larger—and much more costly—PV array to power them.

Well-designed, energy-efficient homes can usually avoid using 240 volts AC. A frequent exception is a deep well pump, which may require 240 volts, but in most cases, a high-efficiency 120-volt AC pump, or even a DC pump, can do the job. Several companies manufacture 120/240 volt inverters, including Magnum Energy, OutBack, and Schneider Electric.

In olden days, installers often stacked inverters to increase the voltage and amperage in battery-based PV systems. Two inverters, for instance, were often stacked to produce 240-volt AC electricity. Those days are quickly coming to an end because inverter manufacturers have created battery-based inverters that produce high voltage without stacking. A single inverter, for instance, can be programmed to produce either 120-volt AC for off-grid applications or 240-volt AC for grid-tied homes (with battery backup). Credit: Outback Power Systems.

Another option—for off-grid homes that require only one 240-volt outlet—is to install a step-up transformer. These units step up the 120-volt AC electricity from an inverter to 240-volt AC electricity.

Conclusion

If you are hiring an honest, experienced professional to install a PV system, you won't have to worry about buying the right inverter. When installing your own system, be sure that your source takes the time to determine which inverter is right for you. Provide as much information about your needs as possible, and ask lots of questions.

A good inverter is key to the success of a renewable energy system, so shop carefully. Size it appropriately. Be sure to consider future electrical needs. But don't forget that you can trim electrical consumption by installing energy-efficient electronic devices and appliances. Efficiency is always cheaper than adding more capacity. When shopping, select the features you want and buy the best inverter you can afford. Although modified square wave inverters work for most applications, it is best to purchase a sine wave inverter.

CHAPTER 7

Batteries, Charge Controllers, and Gen-sets

Batteries are the unsung heroes of off-grid PV systems. Even though PV modules receive most of the attention, work hard, and endure extreme weather, the batteries of off-grid systems operate day in and day out, 365 days a year for up to ten years, give or take a little, provided you treat them well. Although batteries in off-grid systems do indeed work hard, these brutes of the renewable energy field require considerable pampering. You can't simply plop them in a battery room, wire them into your system, and then go about your business expecting your batteries to perform at 100% of their capacity for the lifetime of your system. You'll need to install them properly, watch over them carefully, and tend to their needs. If you don't, they'll die young, many years before their time.

Because batteries are so important to off-grid renewable energy systems and require so much attention, it is essential that readers who are thinking about installing an off-grid system or a grid-connected system with battery backup understand a fair amount about batteries. Doing so will help you get the most from them and could save you a fortune over the long haul. This chapter will also help you develop a solid understanding of batteries and two additional components required in battery-based PV systems: charge controllers and generators. They are vital to the health and vitality of a battery bank.

If you are only interested in a batteryless grid-connected system, you don't need to read this chapter. Feel free to skip to Chapter 8.

Understanding Lead-acid Batteries

Batteries are a mystery to many people. How they work, why they fail, and what makes one type different from another are topics that can boggle the mind.

Fortunately, the batteries used in most off-grid renewable energy systems are pretty much the same: they're deep-cycle, flooded lead-acid batteries.

Flooded lead-acid batteries for renewable energy systems are the ultimate in rechargeable batteries. They can be charged and discharged (cycled) a thousand or more times before they wear out, provided you take good care of them.

Lead-acid batteries used in most renewable energy systems contain three cells— that is, three distinct 2-volt cells or compartments. Inside the battery case, the individual cells are electrically connected (wired in series). As a result, they collectively produce 6-volt electricity.

Inside each cell in a flooded lead-acid battery is a series of thick, parallel lead plates, as shown in Figure 7.1. Each cell is filled with battery fluid, hence the term "flooded." Battery fluid consists of 70% distilled water and 30% sulfuric acid. A partition wall separates each cell, so that fluid cannot flow from one cell to the next. The cells are enclosed in a heavy-duty plastic case.

As illustrated in Figure 7.1, two types of plates are found inside a battery: positive and negative. They are wired in series inside the battery and connected to the

The Life of a Battery

Off-grid systems require batteries. Without them, there would be no electricity during the periods when a PV array is idle. Batteries ensure a steady supply of electricity to off-gridders. When the Sun is not shining or when demand exceeds output of the PV array, batteries are there to provide the electricity needed to keep homes and offices running smoothly. When surplus electricity is available, the batteries store it for later use.

While batteries in an off-grid system work day and night, batteries in a grid-connected system live a life of leisure. Their main function is to store electrical energy in case of an emergency. In between power outages, however, the batteries lounge in the comfort of the battery room, filled to capacity with electricity, which is as good as it gets for a battery. Their sole purpose is to remain fully charged in case they are called into duty.

As a general rule, the largest (and most costly) battery banks are required for off-grid systems. Wind/PV hybrid systems require slightly smaller battery banks, for reasons explained in Chapter 5. Grid-connected PV systems with battery backup generally require even smaller battery banks—typically just enough to get a family or business by for a few hours or a few days should grid power fail.

battery posts or terminals. Battery posts allow electricity to flow into and out of a battery.

As shown in Figure 7.1, the positive plates of lead-acid batteries are made from lead dioxide (PbO_2). The negative plates are made from pure lead. Sulfuric acid fills the spaces between the plates and is referred to as the *electrolyte*. It plays an extremely important role in batteries, and will be discussed shortly.

How Lead-acid Batteries Work

Although battery chemistry can be a bit daunting to those who have never studied the subject or had it explained well, it's important that all users understand a little about the chemical reactions.

Like all other types of batteries, lead-acid batteries convert electrical energy into chemical energy when charging. When discharging (that is, giving *off* electricity), chemical energy is converted back into electricity. How does this occur?

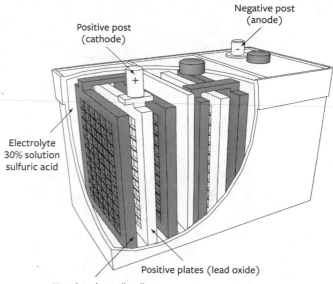

FIGURE 7.1. Cross Section of Lead-acid Battery. Credit: Anil Rao.

As shown in Figure 7.2, when electricity is drawn from a lead-acid battery, sulfuric acid reacts chemically with the lead of the negative plates. As noted in the top equation, this chemical reaction yields electrons, tiny negatively charged subatomic particles (Chapter 1). The electrons flow out of the battery via the negative terminal through the battery cable.

During this chemical reaction, you will note that lead reacts with sulfate ions to form lead sulfate. As a result, tiny lead sulfate crystals form on the surface of the negative plates when the battery is discharging.

The second equation in Figure 7.2 shows the chemical reaction that takes place on the surface of the positive plates when a battery is discharging; sulfuric acid reacts with the lead dioxide of the positive plates, forming tiny lead sulfate crystals on them as well. Keep that in mind, because it's the formation of lead sulfate that reduces a battery's lifespan. Excessive buildup can destroy a battery. Your job as an off-grid solar system operator is to see to it that lead sulfate is quickly removed

FIGURE 7.2. Chemical Reactions in a Lead-acid Battery. The chemical reactions occurring at the positive and negative plates are shown here. Note that these reactions are reversible. Lead sulfate forms on both the positive and negative plates during discharge. Credit: Anil Rao.

$Pb(s) + HSO_4^-(aq) \rightarrow PbSO_4(s) + H^+(aq) + 2e^-$	negative plates
$PbO_2(s) + HSO_4^-(aq) + 3H^+(aq) + 2e^- \rightarrow PbSO_4(s) + 2H_2O(1)$	positive plates

from the plates; you do it by reversing the chemical reaction—that is, charging the batteries as soon as possible after discharging.

Discharging a battery not only creates lead sulfate crystals, it converts some of the sulfuric acid into water. However, as you shall soon see, the chemical reactions just described are reversible. Therefore, when a battery is charged (when electricity is delivered to the battery), the chemical reactions at the positive and negative plates run in reverse. As a result, lead sulfate crystals on the surface of both the positive and negative plates are broken down and sulfuric acid is regenerated. The plates are restored, although a very small number of lead sulfate crystals flake off, slowly but surely whittling away at the plates.

Although the chemistry of lead-acid batteries is complicated, what is most important to remember is that these chemical reactions allow us to store electricity (electrons) inside batteries and reclaim them when we need electricity.

A Word of Warning

Sulfuric acid is a *very* strong acid. In fact, it is one of the strongest acids known to science. In flooded lead-acid batteries, sulfuric acid is diluted to 30%. Although diluted, it is still to be treated with great respect—it can burn your skin and eyes and eat through clothing like a ravenous moth. Be sure to wear eye protection and rubber gloves when filling batteries. It's a good idea to remove jewelry and wear a long-sleeved shirt you don't care about. It's bound to get a little acid on it; and the acid will produce tiny holes in the shirt that you won't notice until you wash it.

Will Any Type of Lead-acid Battery Work?

Lead-acid batteries are used in a wide range of applications. For example, there's a 12-volt flooded lead-acid battery under the hood of your automobile. Trucks and buses have similar batteries, only larger. In all of these vehicles, batteries provide the electricity required to start the engine and operate lights, the ignition system, and accessories like a clock or a radio when the engine's not running. As a result, these batteries are known as *starting, lighting, and ignition*, or *SLI*, batteries.

Lead-acid batteries are also used in a wide assortment of electric vehicles, including forklifts, golf carts, and electric lawn mowers (Figure 7.3). They are even used in emergency standby power systems, providing backup power in case the electrical grid goes down. Lead-acid batteries are also used in RVs, sailboats, yachts, and powerboats, although sealed batteries are typically used in these applications so they won't spill acid when the boat or RV rocks. (More on sealed batteries shortly.)

Lead-acid batteries come in many varieties, each one designed for a specific application. Car batteries, for example, are designed for use in cars, light trucks, and vans. Their thin, porous plates offer lots of surface area to the electrolyte. This allows the battery to discharge large numbers of electrons when you start your car.

This high-amp current enables the battery to turn the starter motor, an event that requires a huge amount of current.

Unfortunately, the thin lead plates of an SLI battery can easily be damaged by deep discharges. Lead sulfate that forms on the plates flakes off, and destroys the thin plates. Leave your car's headlights on a few times and you'll destroy the battery. Marine batteries are similar but are optimized for use in boats—for starting engines and providing small amounts of electricity to power electrical equipment like radios and GPS units. That said, there are deep-cycle marine batteries on the market, too. I'll discuss them shortly.

While most lead-acid batteries—even car batteries—can be installed in a solar electric system, for optimum performance and years of trouble-free service, I recommend using batteries designed specifically for renewable energy systems: deep-cycle, flooded lead-acid batteries (Figure 7.4). They can be safely deep discharged over and over again with relative impunity, for reasons explained shortly.

Even though batteries designed for deep discharge are considerably more robust than car and truck batteries, they are not invincible. If a battery is discharged too deeply—more than 80% of its capacity—it will be permanently damaged. Fortunately, controls in a well-designed renewable energy system prevent this from occurring.

For optimum performance, deep-cycle batteries also need to be recharged fairly soon after undergoing a deep discharge—even if they are only discharged

FIGURE 7.3. My Electric Truck. A couple friends and I converted this Chevy S-10 to operate on electricity. We removed the gas engine and all internal combustion-related components, like the exhaust system and gas tank, and replaced them with an electric motor and 24 deep-cycle golf cart batteries. This truck can travel a little under 60 miles on a single charge and easily cruises at 55 to 60 miles per hour (88 to 97 kilometers per hour). Credit: Dan Chiras.

FIGURE 7.4. Lead-acid Batteries for RE System. Lead-acid batteries are commonly used in renewable energy systems. Be sure to purchase deep-cycle batteries; keep them in a warm, safe place; recharge quickly after deep discharges; and equalize them periodically. Credit: Surrette Battery Company LTD.

20% to 30% of their capacity. The reason for this is that if the batteries are left in a state of discharge for an extended period, the lead sulfate crystals begin to coalesce. Small crystals become large. These crystals "insulate" the electrolyte from the lead plates. This reduces the ability of the battery to store electricity. Because batteries take progressively less charge, they have less to give back. Making matters worse, large crystals are also not completely converted back to lead or lead dioxide when you charge a battery that's been left in a state of deep discharge for an extended period. Over time, then, entire cells in a battery may die, effectively ending a battery's useful life.

To avoid lead sulfate crystal growth, batteries should be promptly recharged and also periodically *equalized*. This process, discussed shortly, helps rejuvenate batteries by driving lead sulfate off the plates. It also equalizes the voltage of the cells in a battery bank.

So, do not forget: *For optimum long-term performance, batteries need to be recharged as quickly as possible after deep discharging to ensure a long, productive life.* Deep-cycle, flooded lead-acid batteries like L16s, should last for seven to ten years, maybe even longer if they are promptly recharged after deep discharging, routinely equalized, and housed in a warm location. If you fail to perform this routine maintenance and make other mistakes like not filling the batteries with distilled water when needed, you can count on having to write a very large check for a replacement battery bank much sooner than necessary.

Can I Use a Forklift, Golf Cart, or Marine Battery?

Forklifts require big, high-capacity, deep-discharge batteries. These leviathans can be used in renewable energy systems because they are designed for a long life and operate under fairly demanding conditions. In fact, they can withstand 1,000 to 2,000 deep discharges—more than standard deep-cycle batteries used in battery-based renewable energy. This makes them ideally suited for renewable energy systems.

Although forklift batteries function very well in such instances, they are rather heavy and expensive. If you can acquire them at a decent price, you might want to use them, especially if the size of your system warrants a large battery. You will, however, probably need to rent a forklift to install them in your battery room.

Golf cart batteries can also work. Like forklift batteries, golf cart batteries are designed for deep discharge. Moreover, they typically cost a lot less than larger deep-cycle batteries. While the lower cost may be appealing, remember that there's a reason for this: golf cart batteries are smaller than standard batteries used in renewable energy systems and, therefore, don't store as much electricity. They also don't last as long. In fact, they may last only five to seven years—if well cared for. Shorter lifespan means more frequent replacement. More frequent replacement means higher long-term costs and more hassle.

Although golf cart batteries don't last as long as the heavier-duty, deep-cycle batteries like L16s, some renewable energy system installers swear by them. Others recommend them for first-time users, who are bound to abuse their first set of batteries. They're a training set. If you screw up and kill your battery bank, you won't be out as much money.

Although marine batteries are advertised as deep-cycle or "marine deep-cycle" batteries, they are really a compromise between a car starting battery and a deep-cycle battery. Their thinner plates just aren't up to the task of a renewable energy system. They will not last as long in deep-cycle service as true deep-cycle batteries.

What About Used Batteries?

Although used batteries can sometimes be purchased for pennies on the dollar, and may seem like a bargain, for the most part, they're not worth it. As a buyer, you have no idea how well—or more likely, how poorly—they've been treated. How old are they? Have they been deeply discharged many times? Have they been left in a state of deep discharge for long periods? Have they been filled with tap water rather than distilled water? Have the plates been exposed to air due to poor maintenance?

Many used batteries are discarded because they've failed or are experiencing a serious decline in function. As noted, with age comes decreased efficiency that translates into decreased storage capacity. If a used battery is 80% efficient, you have to put 120 amp-hours into the battery for every 100 amp-hours you take out. If the battery is only 50% efficient, you have to put 200 amp-hours into it for every 100 amp-hours you draw out.

There are some exceptions to my advice not to buy used batteries. You may be able to find high-quality gently used batteries from telephone companies. They routinely replace batteries in their repeater stations (stations for landlines that amplify signals and transmit them on to the next station).

Whatever your source of used batteries is, shop carefully. Find out how the batteries were used. And remember, although there are exceptions, most people who've purchased used flooded lead-acid batteries have been disappointed.

Sealed Batteries

While most installers of off-grid PV systems recommend flooded lead-acid batteries, they often prefer to install "sealed" lead-acid batteries in grid-connected

Don't Skimp on Batteries

When shopping for batteries for a renewable energy system, I recommend that you buy high-quality deep-cycle batteries. L16s are an industry standard. Although you might be able to save some money by purchasing a cheaper alternative, frequent replacement is a pain in the neck and can be costly. Batteries are heavy, and it takes quite a lot of time to disconnect old batteries and rewire new ones. Bottom line: the longer a battery will last—because it's the right battery for the job and it's well made and well cared for—the better.

FIGURE 7.5. Sealed Lead-acid Battery. Sealed batteries like these 12-volt batteries never need to be watered. Nonetheless, they still require proper housing and temperature conditions and careful control of state of charge to ensure a long life. Notice the battery posts and nonremovable pressure-release caps. Credit: Dan Chiras.

systems with battery backup. Sealed batteries are also known as *captive electrolyte batteries.* They are filled with electrolyte at the factory, charged, and then permanently sealed. Because they are sealed, they are easy to handle and ship without fear of spillage. They won't even leak if the battery casing is cracked open, and they can be installed in any orientation—even on their sides. If you are pressed for space in your battery room, these batteries might work well.

Unlike flooded lead-acid batteries, which require periodic filling with distilled water, sealed batteries never need watering. In fact, you *can't* add distilled water. There are no fill-caps (Figure 7.5).

Two types of "sealed" batteries are available: *absorbed glass mat* (AGM) batteries and *gel cell batteries.* In absorbed glass mat batteries, thin absorbent fiberglass mats are placed between the lead plates. The mats consist of a network of tiny pores that immobilize the battery acid—so it won't spill out if a battery is carried, jostled (as in a sailboat), or laid on its side. The fiberglass mats also create tiny pockets that capture hydrogen and oxygen gases given off by the lead plates during charging. Hydrogen comes from two sources. First is the chemical reactions that occur during charging. Second is hydrolysis,

Sealed Batteries—A Misnomer?

The truth be known, sealed batteries are not totally sealed. Each battery contains a pressure-release valve that allows gases and fluid to escape if a battery is accidentally overcharged. The valve keeps the battery from exploding. Once the valve has blown, though, the battery will very likely need to be retired.

the breakdown of water. As a battery charges, some of the electricity splits water molecules into hydrogen and oxygen. Hydrolysis also produces oxygen. In flooded lead-acid batteries, these gases escape through small openings in the battery caps. It's because of hydrolysis that you need to add distilled water every month to a flooded lead-acid battery.

In sealed batteries, hydrogen and oxygen cannot escape. There are no caps to allow it to vent. These gases accumulate in the tiny pockets in the glass matt where they react with each other, forming water.

In gel batteries, the sulfuric acid electrolyte is converted to a substance much like hardened Jell-O by the addition of a small amount of silica gel. The gel-like substance fills the spaces between the lead plates and eliminates liquid electrolyte.

Sealed batteries are often referred to as "maintenance-free" batteries because fluid levels never need to be checked and because the batteries never need to be filled with distilled water. They also never need to be (and should not be!) equalized, a process discussed later in this chapter. Eliminating routine maintenance saves a lot of time and energy. It makes sealed batteries a good choice for off-grid systems in remote locations where routine maintenance is problematic—for example, back-woods cabins or cottages that are only occasionally occupied. Sealed batteries are also used on sailboats and RVs where the rocking motion could spill the sulfuric acid contained in flooded lead-acid batteries. In these applications, space is limited and batteries are frequently crammed into out-of-the way locations. Sealed batteries are also used in grid-tied systems with battery backup. So, these batteries tend to be ignored—out of sight, out of mind. However, if the batteries are deeply discharged for long periods, they can be ruined.

Sealed batteries offer several additional advantages over flooded lead-acid batteries. One advantage is that they charge faster. Sealed batteries also release no explosive gases, so there's no need to vent battery rooms or battery boxes where they're stored. In addition, sealed batteries are much more tolerant of low temperatures. They can even handle occasional freezing, although this is never recommended. Sealed batteries also experience a lower rate of self-discharge. That is, they discharge more slowly than flooded lead-acid batteries when not in use. (All batteries self-discharge when not in use.)

Sealed batteries do have a few significant disadvantages. One of them is that they are more expensive than flooded lead-acid batteries. Another is that they store less electricity than flooded lead-acid batteries of the same size. Sealed batteries have a shorter lifespan than flooded lead-acid batteries. In addition, they can't be rejuvenated (equalized) like a flooded lead-acid battery. Equalization is a controlled overcharge that produces a tremendous amount of hydrogen and oxygen gas. While safe in flooded lead-acid batteries (because the gases can escape through the caps) equalization in a sealed battery would result in excessive pressure buildup inside the battery. If they were accidentally equalized, pressure would be released through the pressure-release valve on the battery; if this happens, electrolyte is lost, which can destroy or seriously decrease the storage capacity of the battery. Finally, sealed batteries must be charged at a lower voltage setting than flooded lead-acid batteries. Because of these problems, most installers working with battery-based systems recommend flooded lead-acid batteries.

Wiring a Battery Bank

As you learned in the last chapter, PV systems can be wired at either 12, 24, or 48 volts. Twelve-volt systems are typically very small, for example, those used to power RVs, sailboats, and cabins. Although some 12-volt systems include inverters to boost the voltage to 120 volts, many of them run entirely off 12-volt DC electricity. To make this happen, you must install all 12-volt DC lights and appliances.

Larger systems, that is, those that power off-grid homes and businesses, are typically wired to produce 24- or 48-volt DC electricity. The low-voltage DC electricity, however, is typically converted to 120 or 120/240 volt AC electricity.

Just like PV modules, batteries can be wired in series (positive to negative) or parallel. Let's begin with series wiring.

Figure 7.6 shows four 6-volt batteries wired in series. Notice how the positive terminal (electrode) of one battery is wired to the negative terminal (electrode) of

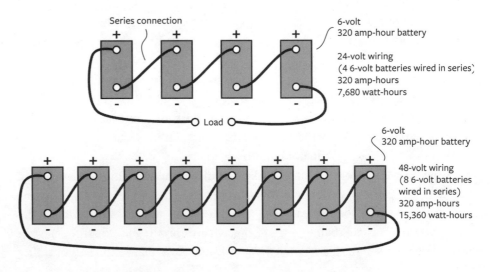

FIGURE 7.6. Battery Wiring: 24- and 48-volt Systems. Batteries are wired in a string in series to boost the voltage. Several strings are wired in parallel. The battery banks here are wired for 24- and 48-volt electricity. Credit: Anil Rao.

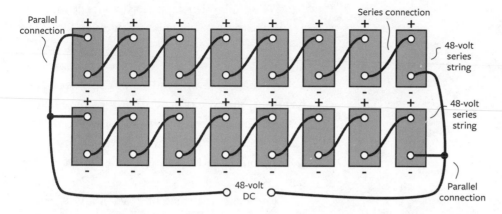

Parallel
connection

Series connection

48-volt
series
string

48-volt
series
string

48-volt
DC

Parallel
connection

FIGURE 7.7. Battery Wiring: Series and Parallel 24- and 48-volt Systems. In this diagram, two 48-volt strings are wired in parallel to boost the number of kilowatt-hours of electricity that are stored and available to a homeowner or business. Credit: Anil Rao.

the next battery in series. Wiring four 6-volt batteries in series produces a 24-volt string. Also notice that each string has an unpaired positive and negative terminal and battery cable. These are commonly referred to as the "most positive" and "most negative." They run to the charge controller. Figure 7.6 also shows a series string comprised of eight 6-volt batteries. Take a moment to study this figure.

While wiring in series boosts the voltage, wiring in parallel increases the amp-hour storage capacity of a battery bank—the number of amp-hours of electricity a battery bank can store. (One amp-hour is one amp of current flowing for one hour—either into or out of a battery bank.) To boost the amp-hour capacity of a renewable energy system, installers typically include two or more series strings of batteries in parallel, as shown in Figure 7.7. As you can see from the diagram, in parallel wiring the positive ends of each series string are connected, as are the negative ends. The most positive and most negative run to the charge controller and inverter.

Lithium Ion Batteries

Lead-acid batteries are the weak link in off-grid and grid-tied systems with battery backup. They are costly, don't store much electricity, and require proper housing (not too hot, not too cold). They also require periodic maintenance and more-frequent-than-you-will-be-happy-with replacement. This has led manufacturers to search for better options. Interestingly, much of the research and development on new batteries is occurring in the electric car industry. Companies such as Tesla, Nissan, and Ford—all leaders in this field—are intensely interested in developing inexpensive, long-lasting, lightweight, and high-capacity batteries to extend the range of their vehicles. In fact, today's electric cars, like my all-electric Nissan Leaf, are all powered by lithium ion batteries (Figure 7.8). Nissan's Leaf has a cruising

FIGURE 7.8. Dan with His Electric Car. The author's all-electric 2013 Nissan Leaf, which he powers entirely with solar and wind energy. Folks interested in buying a used all-electric car can get some killer deals on slightly used vehicles. This car costs over $35,000 new. He got it in January 2016 for under $9,000, and it was in perfect condition. Credit: Linda Stuart.

range of 84 miles (if you purchase a vehicle with the 24-kWh battery) or 107 miles (with the 30-kWh battery). Both are perfect for most commuters. Tesla's Model S, a roomy sedan, travels about 270 miles on a single charge. How far you can go depends on how you drive, the type of road (gravel vs. pavement), road conditions (rain, snow, ice, or clear sailing), terrain (hills vs. flat), use of heaters or air conditioning, outdoor temperature, and the age of the battery.

Elon Musk, chairman of Tesla and First Solar, made headlines in 2015 when he announced a compact, wall-mounted lithium ion battery for household use, ostensibly as a backup for solar customers. My email inbox was soon flooded with questions from students, friends, and colleagues asking whether we've finally got a new battery to replace those clunky lead-acid batteries.

While exciting news, Tesla's Powerwall battery isn't quite what solar homeowners really need—at least not yet. However, in January 2016, the company announced that they would be releasing a newer, better model late in the summer of 2016. So stay tuned.

At this writing (February 2016), Tesla currently offers two batteries, one with a 7-kWh capacity and another with 10-kWh capacity. For reference, a typical lead-acid battery bank in a 48-volt off-grid solar system would store over 50 kWh of electricity; however, to protect batteries and ensure long life, we don't like homeowners to use more than half of that, so a typical battery bank in an off-grid home really only stores 25 kWh of electricity. Clearly, neither Tesla battery comes close to the requirements of an off-grid home. But what about use in a home powered by a grid-tied system with battery backup?

Here the news is a bit more encouraging. In a system designed to meet critical loads—grid-tied with battery backup—a lead-acid battery bank would typically store about 18 kWh (eight 6-volt L16 lead-acid batteries that store about 360–370 amp-hours each). Remember, however, only half of this electricity is available for routine use. So, let's say these battery banks hold about 9 kWh of electricity.

In homes such as this, where the battery bank powers critical loads during a power outage, the 7-kWh and 10-kWh battery might work. The 7-kWh battery, which is designed for daily use, would be a bit too small to power critical loads during a three- to four-day power outage. However, two or more Tesla batteries can be wired together to increase battery storage capacity. According to Tesla's website, the battery can produce 3,300 watts of continuous power. This would very likely not be sufficient for an all-electric home.

One additional consideration: Although the Tesla Powerwall battery looks sleek in photographs, it weighs 210 pounds (96 kg). It is also rather large, measuring 51.2″ × 33.9″ × 7.1″ (1300 mm × 860 mm × 180 mm)—that's over 4 feet by nearly 3 feet. Wall mounting would require a decent amount of space and special precautions to anchor the unit in place.

For most grid-tied systems with battery backup, Tesla's 10-kWh battery would be a better choice. According to the company's website, these batteries produce about 3,300 watts of continuous power. That would limit the number of loads that could be run at any one time. However, for a family that's tuned into energy savings, especially during times of power outages, this battery would probably work. But what would it cost?

In January 2016, the 7-kWh and 10-kWh Powerwalls were listed on Tesla Motors' website at $3,000 and $3,700, respectively, for installers. For comparison, a lead-acid battery bank with eight L16s would cost about $3,200 plus installation. Let's say $5,000. It would last 7 to 10 years. Powerwalls should last about the same length of time.

Tesla's market for its battery is for grid-tied customers who want to add storage so they can use surplus the electricity they produce during the day at night. On their website, they note "each Powerwall has a 7-kWh energy storage capacity, sufficient to power most homes during the evening using electricity generated by solar panels during the day. Multiple batteries may be installed together for homes with greater energy needs. A 10 kWh weekly cycle version is available for backup applications."

Problem is, most homes in America consume at least 1,000 watts of continuous power—that's a kWh every hour. Although most people sleep at night and require less energy at night than during the day, the battery bank would very likely not be sufficient to meet all needs for many households. If you prepare and consume your dinner after dark, watch several hours of TV or have two or three flat-screen TVs operating at the same time, have several computers running at the same time, or wash or dry clothes, the battery won't be able to keep up with your demand. If the weather's cold and you are running a couple of space heaters, which use over 1000 watts each, you are out of luck. If you are running a central air conditioner at night and watching TV, you won't have enough energy to make it through the night.

As you can see, it pays to understand battery capacity, power consumption (watts), and energy usage (kWh used per day) to make a sensible decision about batteries.

Pros and Cons of Lithium Batteries

Manufacturers are currently developing several types of lithium batteries. The safest and most reliable appear to be the lithium iron phosphate batteries. The batteries can undergo many deep discharges. They also charge faster than lead-acid batteries because they can take the full charger output throughout the charge cycle. And, they don't suffer from sulfation, either.

That said, lithium iron batteries require a battery charge management system that balances the charge on a per-cell basis and protects the battery from overheating. Lithium ion batteries also deteriorate over time, at least in cars, but probably when serving as solar backup. Many Nissan Leaf owners have found that the capacity of their battery banks declines each year. (It's a common complaint against what is otherwise a brilliantly designed and well-built vehicle.)

Lithium ion batteries may also require special care to ensure longevity. In their electric cars, for instance, Nissan recommends that the batteries not be charged to more than 80% of their full capacity to ensure longer life. Of course, if you need to charge up for a 78-mile trip, it's okay to charge the batteries to full capacity. Just don't do it every night, as owners are often wont to do. The company also recommends charging at 240 volts, rather than 120. And, it recommends letting the battery bank cool down a bit before plugging in the charger for longer battery life.

Lithium ion batteries also don't function as well in extreme temperatures, although the temperature range in which they function well is much wider than lead-acid batteries. (That's why they work in cars in temperate climates.) Lithium ion batteries also perform less well when it's cold, which is one reason why Nissan Leaf owners see their total range drop over the winter (the other reason is that snowy, icy roads create more friction that requires more energy to overcome).

So, while there are promising developments in the battery world, we still have a way to go.

Sizing a Battery Bank

Properly sizing a battery bank is critical to designing a reliable off-grid PV system or grid-tied system with battery backup. The principal goal when sizing a battery bank for off-grid systems is to install a sufficient number of batteries to carry your household or business through periods when the Sun is unavailable. These are called *battery days*, discussed shortly. In grid-tied systems with battery backup, the same goal applies; however, these battery banks are typically smaller because they only support critical loads.

In off-grid systems, the easiest way to size a battery bank is to calculate how much electricity you use in a day. Because electrical consumption varies from month to month, it is best to use a daily rate from the month with the greatest consumption. In many locations, this occurs during the dead of winter, the least sunny time of year. People spend more time indoors. Lights are on more, as are TVs and computers.

Once you determine how many kilowatt-hours of electricity your home or office will consume on average during a typical day during your most energy-intensive month, you must adjust for the number of sunless days (battery days) that occur during that time. That is, you must determine how long your renewable energy system will be partially sidelined by low irradiance. In sunnier areas, such as Colorado and Wyoming, renewable energy designers plan for three battery days. In even sunnier areas like Arizona, you might be fine with one or two battery days. In the Midwest, five battery days are recommended. (Bear in mind that even on cloudy days, solar electric modules continue to produce electricity, albeit at a reduced rate.) Shorter periods are generally required for those who install hybrid systems—for example, wind and solar electric systems. Fewer battery days may also be appropriate for people who run a backup fossil-fuel generator during cloudy periods. If you're interested in learning more about battery banks, read the accompanying sidebar, "More on Sizing a Battery Bank."

Reducing the Size of Battery Banks

Because batteries are expensive, require periodic maintenance, and take up a lot of room, many off-grid homesteaders install hybrid systems—systems that frequently combine a PV system and a wind generator. This allows them to reduce the size of their battery banks. Most homeowners add a gen-set—a gasoline, diesel, propane, or natural gas generator—to provide additional backup power. Or they increase the size of their PV array. Because PV modules will likely outlast four or five battery banks, the investment in additional generating capacity may be well worth it in the long run. In addition, PV systems require very little maintenance.

Home Power's Ian Woofenden says, however: "Although I love the theory of investing in RE capacity instead of lead (in batteries), there is a limit to its potential."

More on Sizing a Battery Bank

To understand how a battery bank is sized, let's look at an example. Let's suppose that you and your family consume 4 kWh (4,000 watt-hours) of electricity a day on average during the most energy-intensive month of the year. If you need five days of battery backup, you will need 20 kWh of storage capacity (5 days × 4 kWh).

However, it's best not to discharge batteries below the 50% mark because deeply discharging batteries reduces their life. So, if you set 50% as your discharge goal, you'll need 40 kWh—or 40,000 watt-hours—of storage capacity.

To determine how many batteries you need to provide this amount of electricity, first find out how much electricity can be stored in the type of batteries you are using. Battery storage capacity is rated in amp-hours. For instance, a 6-volt L16 deep-cycle flooded lead-acid battery is rated at 360 amp-hours. But how do you know how many amp-hours of storage capacity you need when all you know is that you need 40 kWh of storage capacity?

The easiest way to figure this out is to convert amp-hours to kilowatt-hours. To make this conversion, simply multiply the voltage of the battery by the amp-storage capacity. This is a simple modification of the WAV equation introduced in Chapter 3: watts = amps × volts (W = A × V). In this case, however, we want to determine watt-hour storage in each battery.

The equation can be modified as follows: watt-hours = voltage × amp-hours). A 360 amp-hour 6-volt L16 battery, for example, stores 2,160 watt-hours or 2.16 kWh (when brand new).

The next task is to see how many batteries you will need to store this much electricity. To store 40 kWh of electricity, you'd need 18.5 batteries (40 kWh divided by 2.16 = 18.5 batteries).

If you wanted to wire your PV system at 48 volts, each string would have to contain eight 6-volt batteries. Two strings would be 16 batteries. Since all series strings must be wired with the same number of batteries, you'd need to add a third string of eight batteries. The result would be a system with 24 batteries consisting of three series strings, each with eight 6-volt batteries.

As already noted, battery storage can become quite expensive. At today's prices (February 2016), L16 batteries are running about $350 each. Unless you have an old set of batteries to trade in, you will very likely also have to pay a core charge, a fee that's tagged on to batteries if a buyer doesn't trade in an older battery at the time or purchase. That could come to $20 per battery. And, there may be shipping. All told, these batteries could easily cost $9,000 plus the costs for battery cables, a battery box, and installation.

In cloudy regions, PVs may not produce enough electricity to make a huge difference during winter months. In such instances, batteries and gen-sets may be a preferred option. Ian should know. He lives in the cloudy Pacific Northwest.

Fortunately, the National Renewable Energy Laboratory (NREL) offers online assistance to people trying to figure out the best combinations. The program, called *HOMER*, is designed to optimize hybrid power systems (to find it, visit analysis.nrel .gov/homer/). "This is a great tool for optimizing the size of various components in a hybrid power system," notes NREL's small-wind expert Jim Green. "It typically shows that adding a backup generator will reduce system cost and that optimum battery bank size may be on the order of 8 to 12 hours of storage, much less than 3 days, which is typically used." He adds: "Of course, the optimization point will change over time as the price of propane and natural gas go up. Even so, the generator run time can be quite low.... This model [HOMER] will be more than most homeowners will want to tackle, but some will be able and willing to use it. Equipment dealers find it to be a useful tool."

Battery Maintenance and Safety

Now that you understand lead-acid batteries, your options, and a little about wiring and sizing a battery bank, it is time to turn to the equally important topics of battery care, maintenance, and safety. Battery care and maintenance are vital to the long-term success of a renewable energy system. Proper maintenance increases the service life of a battery. Because batteries are expensive, longer service life results in lower operating costs and less environmental impact over the long haul. Put another way, the longer your batteries last, the cheaper your electricity will be and the less impact you will have on the air, water, and soil. In this section, we'll examine battery care, maintenance, and some safety issues.

Keep Them Warm

Batteries may be the workhorses of a renewable energy system, but they like to be kept warm—and full. Cold conditions dramatically reduce their capacity—the amount of electricity they'll store (Figure 7.9).

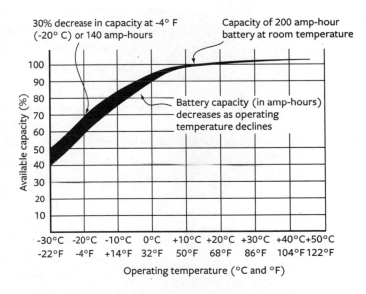

FIGURE 7.9. Battery Performance at Various Temperatures. Batteries function optimally above 50°F. Battery capacity decreases rather dramatically as temperatures in the battery room fall. Credit: Anil Rao.

That's because low temperatures slow down the chemical reactions in batteries, reducing electrical storage.

While low temperatures reduce battery efficiency, higher temperatures result in an increase in outgassing. Outgassing is the release of hydrogen and oxygen gas. It occurs primarily as a result of electrolysis, as just noted. Outgassing not only produces explosive gases, it also reduces battery fluid levels. Outgassing is not a trivial matter. Despite what you might think, Arizona is a harsher climate for car batteries than Minnesota. Higher temperatures also lead to higher rates of self-discharge.

For optimal function, batteries should be kept at around 75°F to 80°F (24°C to 27°C)—about the same temperatures that we like. In this range, batteries will store and hence release a lot more electricity. Guaranteed!

If you can't ensure this narrow temperature range, shoot for a range between 50°F and 80°F (10°C to 27°C). Rarely should batteries fall below 40°F or exceed 100°F—that's 4.5°C and 38°C.

Ideally, batteries should be stored in a separate battery room or a battery box inside a conditioned space held at the optimum temperature. Basements make good candidates for battery rooms or battery boxes. They tend to range between 60°F (16°C) in the winter and 70°F (21°C) in the summer. Heated and cooled shops or utility rooms may also work. I discourage people from housing batteries inside their homes, unless they can build a separate room. I housed mine in Colorado in a well-sealed, ventilated, specially built closet in my utility room.

Whatever you do, don't store batteries in a cold buildings—for example, garages, barns, or sheds. Expect very short service lives when batteries are kept cold. Furthermore, when batteries are in a low state of charge, they can freeze more easily. Freezing causes the water inside a battery to expand and can crack the plastic cases, allowing sulfuric acid to leak out, creating a dangerous mess in the battery room. (Deeply discharged batteries are especially prone to freezing because the sulfuric acid has reacted with lead and lead dioxide to form lead sulfate, leaving a higher concentration of water, which makes the battery more prone to freezing.) As just noted, extremely hot environments are also not recommended.

Batteries should not be stored on concrete floors. Cold floors cool them down and reduce their capacity. Always raise batteries off the floor, so they stay warmer.

If you must store batteries outside your home, be sure to heat and cool the building. Heating and cooling a shed, garage, or barn can be quite expensive and waste valuable fossil-fuel energy. Plus, it can gobble up all the energy a PV system generates, perhaps even more. As an example, I troubleshot a grid-tied system with battery backup installed by another installer. The homeowners were irate because their winter electric bill was higher *after* they installed their PV system. When I visited the site, I found out that they installed a portable electric heater in their detached garage where the batteries were housed, and ran it all winter long. They did this on the advice of the installer. He's told them they'd need to keep the batteries warm for optimum performance. Unfortunately, electric heaters consume tons of electricity—more than a medium-sized solar system can produce.

If you must heat a battery room, insulate it and retrofit it for passive solar by installing south-facing windows with window shades that can be closed at night. This will allow the low-angled winter sun to warm the building. A solar hot air collector might also help keep a garage or shed warmer. Insulation will help keep the building warmer in the winter and cooler in the summer. (For more on passive solar design, see my book *The Solar House: Passive Heating and Cooling*. For information on solar hot air collectors, check out my book *Homeowner's Guide to Renewable Energy*.)

Ventilating Flooded Lead-acid Batteries

To ensure safe operation, battery boxes must be ventilated to the outside to remove potentially explosive hydrogen and oxygen gases released when flooded batteries charge. (This does not apply to sealed batteries.) A small 2-inch (5-cm) vent made from PVC or ABS pipe is all that's typically required to vent a battery box. It should be placed high, as hydrogen is fairly light and rapidly rises to the top of the box.

Battery rooms may also require venting (Figure 7.10). However, hydrogen disperses fairly rapidly in a room and requires very little air movement to prevent

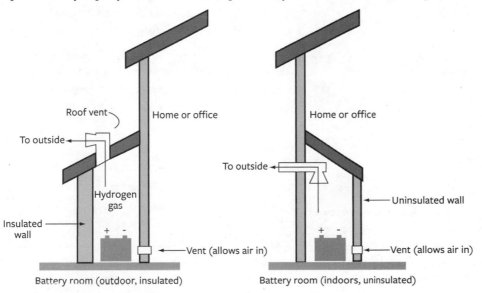

FIGURE 7.10. Battery Vent System. Flooded lead-acid batteries require venting to remove hydrogen released when batteries are charging from a PV array, gen-set, or grid. Vents drawn here are supersized to emphasize their importance. In most cases, smaller passive vents are required. Be sure to allow replacement air to enter near the bottom of the battery room.

it from building up to dangerous levels. Natural air movement in a battery room combined with normal air changes that occur in occupied spaces—or as a result of heating systems—are typically sufficient to disperse oxygen and hydrogen gases generated in most battery rooms. It all depends on the size of the room. The larger the room, the less likely venting will be required.

When venting a battery box or a battery room, be sure that the vent system doesn't cool the space in the winter. Place the opening on the downwind side of the building. Also, be sure that appliances and devices that require an open flame (a gas water heater, for example) or might produce sparks (an arc welder, for example) are not housed near a battery bank.

As illustrated in Figure 7.10, proper ventilation requires an air outlet near the top of the box or ceiling of a battery room to vent hydrogen gas to the outside. It also requires small air inlets near floor level. These openings allow fresh air into the battery box or battery room. Check with your building department to be sure you comply with their requirements, if any.

Battery rooms are generally passively ventilated. However, they can also be actively vented or power vented—for example, if the pipe run is long and/or contorted (not a straight run to the outside). Power venting requires a small electric fan that exhausts hydrogen gas while batteries are charging.

As nifty as power venting may seem, it's generally not necessary. Hydrogen is extremely light and easily escapes if a room is properly vented. So unless the vent pipe is long and contorted or you live in a cold climate, you probably don't need to power vent a battery bank. If, however, batteries are located in a room used for other purposes, for example, a shop or garage, active venting may be prudent. It reduces the chance of explosions.

Battery Boxes

Battery boxes are typically built from plywood (Figure 7.11). Some individuals build them with wooden lids; others incorporate clear plastic (polycarbonate) in their lids. Be sure to seal the lids to prevent hydrogen from escaping. And, be sure to line the boxes with an acid-resistant liner to contain possible acid spills from cracked battery cases. I use black

FIGURE 7.11. Battery Box. We made this battery box from plywood. Be sure to vent and seal the battery boxes when installing flooded lead-acid batteries just in case a battery leaks. This photo shows the ABS plastic lining we caulked in place to seal the battery box to prevent potential leakage. Credit: Dan Chiras.

ABS plastic to line battery boxes. I purchase it in 4 × 8 sheets, cut it with a circular saw (fine-tooth saw blade), and glue it in place with silicone caulk. I then caulk all the joints. Don't use polyethylene plastic. Although it is cheaper than ABS, it won't hold silicone caulk. As an alternative, battery leakage can be prevented by installing battery boots on each battery or by placing them in plastic tubs (Figure 7.12). Choose heavy-duty polyethylene tubs.

Lids should be hinged and sloped to discourage people from piling items on top. The steeper the lid, the better.

As noted above, be sure to place battery boxes in a warm room and always raise boxes off cold concrete floors. I install 2 × 4s under my battery boxes to keep them off concrete floors. Also be sure that the boxes or shelves you build are strong enough to support the weight of the batteries. Large solar batteries weigh 80 to 125 pounds each—that's 36 to 57 kg. Also, be sure to install a lock on battery boxes to keep kids out.

Battery storage boxes can be purchased from commercial outlets. Radiant Solar Tech, at radiantsolartech.com, for example, manufactures sealed plastic battery boxes with removable lids. This company will also custom build boxes for homeowners.

When shopping for battery boxes, note that many commercially available products are designed for sealed batteries, not flooded lead-acid batteries. Before you buy a manufactured battery storage cabinet, be sure that it will work with the batteries you are planning on using in your system. Cabinets for flooded lead-acid battery banks should provide sufficient clearance to view electrolyte levels and to fill cells. Because sealed batteries don't require watering, the shelves are usually spaced very close together, which would make it impossible to service flooded lead-acid batteries—that is, check electrolyte levels and fill the batteries with distilled water.

Finally, batteries should be located as close to the inverter and charge controller as possible. These wire runs require very large-diameter wire, which is expensive. Minimizing the distance saves money, and it minimizes voltage drop that will help make your system more efficient.

FIGURE 7.12. Battery Boots and Plastic Tubs. I installed individual battery boots on the batteries in my off-grid home in Colorado to contain possible leaks. Durable plastic tubs can also be used. Don't buy cheap ones. Batteries are heavy and could cause tubs to crack. Credit: Dan Chiras.

Keep Kids Out

I mentioned it earlier, but it bears repeating: battery rooms and battery boxes should be locked—*especially* if young children are likely to be about. This will prevent kids from coming in contact with the batteries and receiving electrical or acid burns.

Managing Charge to Ensure Longer Battery Life

Housing flooded lead-acid batteries at the proper temperature and keeping them topped off with distilled water ensures a long life. Longevity can also be ensured by proper charge management—keeping batteries as fully charged as possible. Battery banks should also be equipped with temperature sensors that connect to battery chargers. They help regulate charging according to room temperature.

To understand why it is so important to manage battery charge, let's take a look at *cell capacity*, the amount of electricity a battery can store.

"Cell capacity is the total amount of electricity that can be drawn from a fully charged battery until it is discharged to a specified battery voltage," according to Richard J. Komp, author of *Practical Photovoltaics: Electricity from Solar Cells*. Battery cell capacity is measured in amp-hours. A Trojan L16H battery (H stands for high-capacity) can store 420 amp-hours of electricity. Theoretically, a battery with a 420-amp-hour storage capacity could deliver one amp of electricity over a 420-hour-period, or 420 amps for one hour. In reality, the amount of electricity that can be drawn from a battery varies with the discharge rate, that is, how fast a battery is discharged. As a rule, the faster a lead-acid battery is discharged, the less you'll get out of it. A lead-acid battery discharged over a 20-hour period, for example, will yield 100% of its rated capacity. Discharge that battery in an hour and a half, and it will deliver only 75% of its rated capacity. Figure 7.13 illustrates this concept.

To standardize the industry, most batteries are rated at a specific discharge rate, usually 20 hours (which is what the 420-amp-hour capacity in the example above is based on).

Like many devices, lead-acid batteries last longer the less you use them. More specifically, the fewer times a battery is deeply

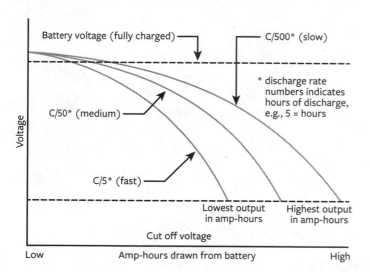

FIGURE 7.13. Discharge Rate vs. Amp-hours Delivered. The rate at which a battery bank is discharged determines how much energy can be obtained. The slower the rate of discharge, the more energy. Credit: Anil Rao.

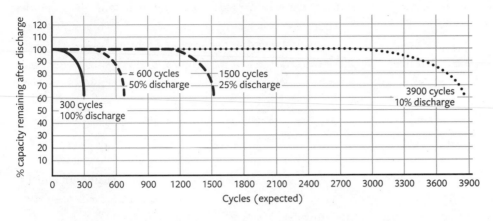

FIGURE 7.14. Battery Life vs. Deep Discharge. This graph shows that shallow discharging results in more cycles. As a result, batteries will last longer if they are not discharged as deeply. Credit: Anil Rao.

discharged, the longer it will last. As illustrated in Figure 7.14, a lead-acid battery that's regularly discharged to 50% will last for slightly more than 600 cycles, if recharged after each deep discharge. If regularly discharged no more than 25% of its rated capacity—and recharged after each discharge—the battery should last about 1,500 cycles. If the battery is regularly discharged only 10% of its capacity, it will last for 3,600 cycles. So, even though deep-cycle batteries can handle deep discharges, shallow cycling makes them last longer.

As Ian Woofenden points out, however, this topic—like so many issues in life—is a bit more complicated. While deep discharging reduces the lifespan of a battery, renewable energy users are more concerned with the cost of the battery per watt-hour cycled. In other words, what we want from batteries is not simply to "last a long time" but to cycle a lot of energy at an affordable price.

Studying the cost of a battery per cycled kWh, says Ian, provides a better measure than total years of service. Theoretically, you'll get the most bang for your buck by cycling in the 40% to 60% discharge range. This goes against the "shallow cycling makes batteries last longer" idea. (After all, if you double the size of a battery bank and reduce the discharge cycle by half, you'd only break even on the larger investment.)

Battery life also depends on frequent recharging of batteries after deep discharges. As you may recall, charging removes lead sulfate from the lead plates. Whatever you do, never leave batteries at a low state of charge for a long time. Lead sulfate crystals will enlarge and could damage your batteries. A good rule to follow is to be sure that the battery bank is fully charged at least once a week (and fully charged does not mean mostly charged!). Think of it this way, if you don't top them off, you start losing the top, starting a downhill slide.

Protecting batteries from deep discharge is easier said than done. As Ian points out, "in real life you often don't have such close control of how deeply your batteries cycle." If your system is small and you don't pay much attention to electrical use, you'll very likely overshoot the 40% to 60% mark time and time again. If you monitor your electrical usage more carefully, and charge batteries shortly after periods of deep discharge, you are more likely to achieve this goal.

One way of reducing deep discharge is to conserve energy and use electricity as efficiently as possible. Conserving energy means not leaving lights or electronic devices running when they're not in use—all the stuff your parents told you when you were a kid. Energy efficiency means installing energy-efficient lighting, appliances, electronics, and so on—the ideas energy conservation experts have been suggesting for decades. In addition, it means eliminating phantom loads.

Energy conservation and energy efficiency are only half the battle, however. In an off-grid system, you will also have to adjust electrical use according to the state of charge of your batteries. In other words, you will need to cut back on electrical usage when batteries are more deeply discharged and shift your use of electricity to sunnier days when the batteries are more fully charged. You may, for instance, run your washing machine and microwave on a bright sunny day when your batteries are full, but hold off during cloudy periods when batteries are running low—unless you want to run a backup generator. Learning how to monitor the state of charge in your batteries is a skill you can learn. It requires a high degree of awareness, however.

Experienced battery operators watch the weather and their battery banks like hawks. If they experience a couple cloudy days, they know that they'll have to cut back on electrical consumption, for example, by cooking on a gas-powered range, as opposed to a microwave.

I monitored my off-grid system in Colorado by checking the battery voltage on the LCD screen on my system at the *end* of each day, when the batteries were no longer charging and demand for power in the house was minimal. The voltage reading gave me a good idea about the state of charge of my batteries. The higher the voltage, the more electricity my batteries held.

Voltage is a surrogate measure of state of charge. That is, it is a good approximation of battery capacity. Remember, however, battery voltage and state of charge are not directly related unless the batteries are at rest. At rest means they are not being charged or discharged or have not been recently charged or discharged. That's because the voltage of batteries that are being charged will read a bit higher than when charging ceases. Technically, a battery is only considered to be at rest when there has been no flow of energy in or out for a period of at least two hours (they need a little time for the voltage to settle). Over the years, however, I've found that it doesn't take very long for a battery voltage to settle down once charging ends.

A more precise way to monitor the state of batteries is with a digital amp-hour or watt-hour meter. One popular meter, the TriMetric, keeps track of the number of amp-hours of electricity moving in and out of a battery continuously. By doing so, and by it knowing the size of the battery in amp-hours, which is programmed into the meter, it can tell you how many amp-hours are stored in the battery at any point in time. To make your life easier, the TriMetric displays a "fuel gauge" reading that shows the state of charge as a percentage, similar to those on laptop computers.

Homeowners should use battery state of charge information to adjust daily activities. If batteries are approaching the 40% to 60% discharge mark, hold off on activities that consume lots of electricity—like vacuuming or toasting bread.

Another way to minimize deep discharging is to oversize a PV system—that is, to install extra modules, or to install a hybrid system—for example, add a wind machine to supplement a PV array. This will ensure that your batteries are supplied with more electricity throughout the day and will lessen chances of deep discharging. Oversizing a battery bank has just the opposite effect.

A backup generator can also be used to manage batteries. During long cloudy periods, for instance, a homeowner can fire up a generator to boost battery voltage. Gen-sets can be wired directly to the inverter or charge controller so they start automatically when the battery voltage drops to a predetermined level. Bear in mind that not all inverters and not all generators operate automatically, so if you want this feature, be sure your generator and inverter are designed for automatic operation. For more on battery charging, see the accompanying sidebar "Understanding Battery Charging."

Watering and Cleaning Batteries

As noted earlier, the lead plates in batteries are immersed in a 30% solution of sulfuric acid. Sulfuric acid participates in reversible chemical reactions that store and release electrons. Because these reactions are reversible, you'd think that battery fluid levels would remain constant over the long haul. Unfortunately, that's not the case. Battery fluid levels decrease over time for three reasons. The first, and most important, is hydrolysis, the breakdown of water by electricity. Second, a small amount of water can evaporate and escape through battery cap vents. Third, sulfuric acid can escape through the vents during charging. It escapes as a result of the strong bubbling action of hydrogen and oxygen gases that occurs during charging. This results in the production of a mist of acid that escapes through battery cap

Understanding Battery Charging

Batteries are charged by the PV array via the charge controller when a PV system is producing electricity. The charge controller performs several key functions. It feeds DC electricity from the array to the battery bank in a highly controlled manner. It also prevents batteries from overcharging. It does this by terminating the flow of DC electricity from the array when the battery bank is full—voltage only climbs to a programmed set point.

As you may recall from Chapter 5, batteries are also charged by the battery charger located in more recent battery-based inverters. The charger converts 120-volt AC electricity from the gen-set or the utility in grid-connected systems with battery backup to DC electricity. It also decreases the voltage so that the DC electricity fed to the battery bank matches the voltage at which it was wired. Converting AC to DC is the function of a rectifier.

Battery charging occurs in three stages—if the batteries have been deeply discharged. These stages are controlled by the charge controller, if the battery bank is being charged by the array. If the battery bank is being charged by the inverter, the battery charger in the inverter exerts its control on charging.

The first stage of battery charging is known as the bulk stage. During this stage, shown on the left in Figure 7.15, the battery charger (or charge controller when the PV array is charging a battery bank) delivers a constant and relatively high charge (lots of amps). As the charging source "pumps" amps into the battery bank, the battery voltage increases. Once the battery voltage reaches a certain point, known as the voltage regulation set point, or bulk volts setting, the bulk stage ends. At the end of the bulk stage, the battery is about 80%–90% charged.

In the second stage, known as the absorption stage, the battery voltage is kept constant and the charge rate (amps) slowly declines. The absorption stage typically lasts about two hours to ensure that a battery is fully charged. Absorption time is determined by the type of battery and is an adjustable setting. When an installer first programs a battery-based inverter, he or she is asked the type of battery, and the inverter then sets the duration of the absorption stage.

At the end of the absorption stage, the charging cycle enters the third and final stage, the float stage. This occurs if the batteries are not being used. During the float stage, the voltage is reduced slightly and held constant at the float volt setting. The current is maintained at a level just a little higher than that which is needed to offset the self-discharge rate of the battery. The float stage is akin to trickle charging a battery, for example, trickle charging a motorcycle battery over the winter to prevent self-discharge that could, over the course of a winter, run a battery dry.

In an off-grid system in which the generator is controlled automatically by the inverter, the generator is switched off as soon as the charger enters the float stage. This minimizes generator run time and fuel consumption. In a grid-connected system, the float stage continues indefinitely with current supplied by the PV array or the utility grid, depending on the time of day and the programmed settings.

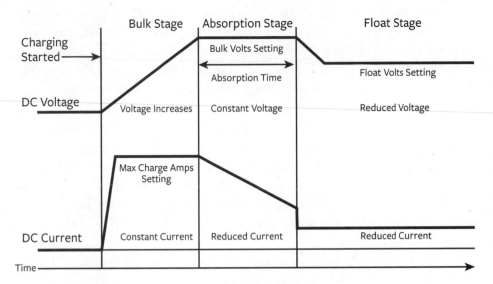

FIGURE 7.15. Battery Charging. Charge controllers and battery chargers in battery-based PV systems carefully control the flow of electricity into a battery bank. For a description, see text. Credit: Apollo Solar.

vents. This mist precipitates out like rain on the top of the battery. Water evaporates, leaving behind a find powder of sulfuric acid.

All of these potential sources of battery fluid loss add up over time and can run a battery dry. When the plates are exposed to air, they quickly begin to corrode. A battery's life is pretty well over when this happens. I have tried all kinds of tricks to bring batteries that have fallen victim to low-electrolyte levels back to life, but to no avail. Even if only one cell in a battery goes bad, the entire battery is out of commission. If a battery bank is more than a year to a year and a half old, you can't replace a bad battery in the system with a new one. They generally all have to go—or, at least that's the advice of battery experts.

To prevent this potentially costly occurrence, monitor battery fluid levels regularly. Many experts recommend checking batteries every month. Others recommend checking batteries every three months. I have found that a monthly checkup works best for my off-grid systems, and two- to three-month checkups work best for my grid-tied system with battery backup. Be sure to note battery checkups on your calendar or enter them into your daily reminders on your cell phone.

Be careful not to get lulled into complacency after installing new batteries, however. The first year of a battery's existence is much like the year or two after a honeymoon. Peace and happiness prevail. That is to say, brand new batteries operate very smoothly for a year or so with very little water loss. As time passes, however, batteries, like our partners in marriage, can become more demanding—a lot more demanding. They require more frequent inspection and watering. And remember, divorce from your battery bank is not an option if you are living off grid. Remember, too, that batteries in hotter climates will require more frequent watering than batteries in cooler climates. In the Dominican Republic, for instance, you might need to water batteries twice a month during the sunniest part of the year. Systems that experience high discharging and charging rates will also require more frequent watering.

To check fluid levels, unscrew the caps and peer into each cell when the system is not charging. Use a flashlight, if necessary—never a flame from a cigarette lighter or any other source! Be sure that the batteries are installed in a battery room or in a battery box in a way that allows them to be easily inspected. Avoid stacking batteries on shelves that preclude you from safely peering into each the cells. That means shelves need to be widely spaced. You'll also need room to fill the batteries, a topic discussed below.

Battery acid should cover the plates at all times—at a bare minimum at least ¼ of an inch above the plates. (That's usually about ¼ inch below the vent or fill tube—the opening in the battery casing into which the battery cap is screwed.) It's better to keep the battery more fully filled, but don't overfill it, either. Overfilling a battery will result in battery acid bubbling out of the cells when batteries are charged. As you just learned, water evaporating from the acid on the top of the battery leaves a white acidic deposit. It not only looks messy, it supposedly can conduct electricity, slowly draining the batteries. (Though I have trouble believing that.) Battery acid also corrodes metal—electrical connections, battery terminals, and battery cables. In addition, the loss of battery acid will result in a dilution of the electrolyte within the battery cells because the addition of distilled water to compensate for lost fluid will continually dilute the remaining acid.

When filling a battery, be sure to add only distilled (or deionized) water. Never use tap water. It may contain minerals or chemicals that could contaminate the battery fluid and reduce a battery's lifespan. Be sure not to overfill batteries.

Batteries should be 100% charged before filling them with distilled water. This occurs on a bright, sunny day when demand is low or nonexistent. You can also fill batteries right after equalization.

Don't fill depleted batteries and then charge them with a backup generator. This could cause battery fluid to bubble out. If the fluid levels are dangerously low, however, add a tiny bit of distilled water, then fully charge the batteries. After that, it is safe to top off the batteries with distilled water.

Battery acid bubbling out of batteries should be cleaned promptly, although most of us don't monitor our batteries that closely. Use paper towels or a clean rag to sop up the liquid. If you don't monitor your batteries to detect spills, be sure to check for white acid powder each time you fill your batteries. Some sources recommend using a solution of baking soda (sodium bicarbonate) to neutralize battery acid that bubbles up onto the top of batteries. After carefully rinsing batteries with sodium bicarbonate, they say, batteries should be wiped down with distilled water again. I discourage people from cleaning batteries with baking soda because it could drip into the cells of a battery, neutralizing the acid and reducing battery capacity. It's been my experience that a damp paper towel or rag is sufficient to clean the surface of batteries. When cleaning batteries, be sure to wear gloves, protective eyewear and a long-sleeved shirt—one you don't care about. If you get acid on your skin, wash it off immediately with soap and water. If you feel a burning itch on exposed skin it means you've got a little acid on you. Wash it off.

Although I don't recommend using baking soda to clean batteries, it is a good idea to keep a few boxes of baking soda near your batteries in case there's an acid spill in the battery room. Acid spills are extremely rare but may occur if a battery case cracks. If one occurs, neutralize the acid with baking soda before mopping it up. You should also install a dry chemical fire extinguisher for safety.

When filling batteries, be sure to take off watches, rings, and other jewelry, especially loose-fitting jewelry. Also empty shirt pockets of cell phones, metal-frame eye glasses, pens, and such. Metal objects such as jewelry will conduct electricity if they contact the positive and negative battery terminals. Such an event will convert your jewelry to a puddle of metal and could cause a serious burn. One 6-volt battery could produce over 3,000 amps if the positive and negative terminals of a battery are shorted. As if that's not enough, sparks could ignite hydrogen and oxygen gas in the vicinity. Heat produced by a short circuit could cause the battery case to explode, ejecting hot battery acid in all directions. You'll surely lose an eye or two if you are not careful.

Also, be careful with tools when working on batteries—for example, when tightening cable connections. A metal tool that makes a connection between oppositely charged terminals on a battery becomes red hot and may be instantaneously welded in place. It can also ignite hydrogen gas, causing an explosion. Short-circuiting a battery will also very likely ruin the battery. To prevent these problems, be sure to wrap hand tools used for battery maintenance in electrical tape so that only one inch of metal is exposed on one end (that way, it can't make an electrical connection). Or, buy insulated tools to prevent this from happening. I use a socket wrench with a stout rubber handle. It saved my life on one occasion when I was working on wiring at the inverter. It is a good idea to have a dedicated set of tools (box wrench or socket wrench) for use only on the battery bank. Find a place for them in a box next to the batteries. That way, you won't be tempted to use whatever (non-insulated) tools might be at hand at the moment you need them.

You may need to clean the battery terminals every year or two. The fine corrosive mist containing sulfuric acid released when a battery charges corrodes the metal posts, battery cables, and hardware. It forms a white/turquoise gooey powder. (The green is from the corrosion of the copper wires in battery cables.) Corrosion increases resistance in the circuit that will reduce battery performance (efficiency) when charging and discharging. Even if corrosion only occurs on one battery terminal, every battery in the series string will be affected. Corrosion is often indicative of a loose connection.

To clean the posts, shut down the system. Disconnect the battery bank from the charge controller. Remove the battery cables that need cleaning. Then use a small wire brush and, if necessary, a spray-on battery cleaner. Cleaners can be purchased at a hardware or auto parts store. Before you reconnect the batteries, be sure to wire-brush the battery terminals so they are shiny. After they are cleaned, reinstall the nuts and bolts, or replace them with new ones.

To reduce maintenance, coat battery posts and ends of the battery cables with a battery protector/sealer, available at hardware and auto supply stores. It's sprayed on the ends of the battery cables and terminals, where it forms a protective layer that halts or greatly reduces corrosion. Some homeowners coat battery terminals with petroleum jelly. Unfortunately, this product may collect dust over time, and it's messy to work with if you need to remove a battery cable or tighten a bolt. Be sure to use flat washers and lock nuts when installing battery cables.

I use flat copper bar stock instead of battery cables. This type of battery connection is easier to make than battery cables, which require you to carefully crimp a terminal on each end. To make a connector with flat copper bar stock, simply cut the material to length and then drill an appropriately sized hole on each end. Before you cut it to length, however, be sure to see if you will need to bend the stock to fit. If so, be sure to cut a bit longer. I find that copper bar stays very clean and presents a lot less corrosion than standard battery cables. I use ⅛-inch thick, ¾-inch wide copper bar stock—it's called Multipurpose 110 copper—from McMaster-Carr, one of my favorite online sources of hardware, fans, and just about everything else I need.

Equalization

Another key to ensuring the longest possible battery life is periodic equalization. *Equalization* is a carefully controlled overcharge of batteries that, among other things, removes lead sulfate from lead plates. If done properly, meaning at the right times of the year, periodic equalization can greatly extend battery life.

Batteries can be equalized in several ways. In some cases, they can be equalized by the PV array. I equalized batteries in my off-grid PV system during the summer using the PV array. It's possible to do this only if it's sunny and the batteries are pretty full. (Unfortunately, summertime equalization is not often needed.)

Batteries can also be equalized with a wind turbine or some other renewable energy source in a hybrid system. A strong storm in the winter or early spring, for example, can provide enough electricity to equalize a battery bank using a wind turbine.

Equalization is most commonly carried out by running a backup generator. Periodic equalization is performed for three reasons: The first is to drive lead sulfate crystals off the lead plates, preventing the formation of larger crystals, which reduce battery capacity, are difficult to remove, and therefore can permanently damage the lead plates.

Batteries are also periodically equalized to stir up (de-stratify) the electrolyte. Sulfuric acid is heavier than water and tends to sink, settling near the bottom of the cells in flooded lead-acid batteries. During equalization, hydrogen and oxygen gases released by the breakdown of water create bubbles that cause the battery to "boil." (This phenomenon is called *boiling* because the rapid formation of bubbles

Sulfate and Sulfation, What's Normal and What's Bad

During discharge in a lead-acid battery, lead sulfate, $PbSO_4$, is formed on the positive and negative plates. When first produced, lead sulfate forms a soft deposit on the plates. If the battery is promptly and fully recharged, lead sulfate is converted back into plate material (lead on the negative plates and lead dioxide on the positive plates) and sulfuric acid (electrolyte). However, if the battery is left discharged, even partially discharged, for some time, the lead sulfate slowly forms large crystals. These crystals are hard and are not easily converted back into plate material during a subsequent charge. An accumulation of these hard crystals is called *sulfation*.

Sulfation is bad for a couple reasons. First, lead sulfate is an insulator. It reduces the chemical reactivity of the plates. Lead sulfate crystals therefore effectively reduce the surface area of the plates. In effect, batteries get smaller and their capacity to store energy is reduced. Second, hard crystals of lead sulfate eventually break off and fall to the bottom of the cell, reducing plate material. Not only is useful plate material lost forever, but the accumulation in the bottom of the cell eventually reaches the bottom of the plates. This prevents electrolyte from contacting the surface of the plates. At this stage, if not before, the battery is so degraded that it must be replaced.

resembles boiling in water; it is not caused by high temperatures.) The bubbles mix the fluid so that the acid is de-stratified in each cell of each battery, which improves efficiency.

Equalization is also performed to help bring all of the cells in a battery bank to the same voltage. That is, it equalizes voltage in the cells of a battery bank. That's important—indeed vital—to a battery bank because some cells may sulfate more than others. As a result, their voltage may be lower than others. A single low-voltage cell in one battery reduces the voltage of the entire string. Batteries and battery banks operate much like a camel train. A camel train operates at the speed of its slowest camel. Similarly, a battery operates at the voltage of its lowest cell. Low voltage in one cell drags the voltage of the entire battery down. One low battery, in turn, drags the voltage of the entire battery bank down.

If equalization restores batteries, why won't a battery remain functional for an eternity? Although equalization removes lead sulfate crystals from plates, restoring their function, some lead dislodges—flakes off—during equalization, settling to the bottom of the batteries. As a result, batteries lose lead over time and never regain their full capacity. This sloughing slowly whittles away at the lead plates. As the amount of lead decreases, the battery's ability to store electricity declines. Lead that sloughs off positive and negative plates also builds up on the bottom of each cell, and can short out a cell.

Equalization is fairly simple. In systems with gen-sets, the owner simply sets the inverter to the equalization mode and then cranks up the generator. The inverter or charge controller controls the process from that point onward. In wind/PV hybrid systems, the operator sets the wind generator charge controller to the equalize setting during a period of high wind. The controller takes over from there.

During equalization, the charge current (number of amps fed into a battery bank) is kept relatively constant and high for a set period, usually four to six hours. During this process, the battery voltage is allowed to rise to higher than normal levels. (Normally the charge controller shuts off or slows the flow of electricity to batteries once a certain voltage is reached.) Over time, the voltage of all of the cells of all the batteries in a battery bank rises to the same level (equalizing them). When the time is up, equalization is terminated automatically.

During equalization, voltage may rise quite high. For a 12-volt system, the voltage may rise as high as 15 to 16 volts. In a 24-volt system, it may rise to 30 to 32 volts. In a 48-volt system, it may rise to 60 volts or higher. Bear in mind that DC loads fed directly from the charge controller that are sensitive to high voltage should be disconnected during equalization. In AC-only systems, the inverter regulates its output voltage to protect loads.

How often batteries need to be equalized depend on how they are used. If your batteries are frequently deeply discharged, or if composed of multiple series strings in parallel, you may want to equalize them once a month. I have found that batteries in my system in Colorado were worked much harder in the late fall, winter, and spring. They were more likely to be deeply discharged, so I equalized more often during this period—every month or two. In the late spring, summer, and early fall, I tended to equalize a lot less frequently, unless we experienced unusually cloudy weather during those periods. If you live in a very sunny location and your batteries are rarely deeply discharged, they'll need less frequent equalization. For example, batteries that are kept pretty full and rarely or never discharged below 50% may only need to be equalized every six months. Needless to say, there are no hard and fast rules. You have to stay tuned to the weather and the demands on your battery bank, and equalize as necessary.

The frequency of equalization also depends on your normal charge set points. Some homeowners run their battery banks "hot"—at a higher voltage setting than the standard. So the batteries have less need of equalization, but more watering.

Rather than second-guess your battery's need for equalization, it is wise to check the voltage of each battery (using a voltmeter) every month or two. If you notice significant differences in battery voltage in a battery bank—that is, the voltage of one or two batteries is substantially lower than others—it is time to equalize the battery bank. Checking voltage requires a voltmeter (Figure 7.16). They are typically referred to as *multimeters* because they are designed to measure a variety of different electrical parameters. A digital multimeter is much easier to read than a dial meter.

Another way to test batteries is to measure the specific gravity of the battery acid using a hydrometer (Figure 7.17). Specific gravity is a scientific term for the density of a fluid. Density is related

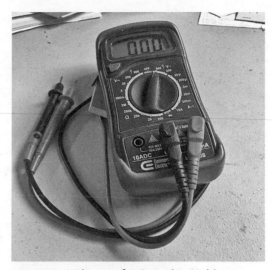

FIGURE 7.16. Voltmeter for Batteries. Multimeters like these can be used to measure the DC voltage of a battery. You can purchase them at hardware stores and home improvement centers as well as electrical supply houses. Don't waste your money on an analog unit. Go digital! Credit: Dan Chiras.

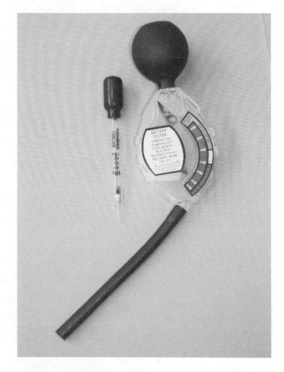

FIGURE 7.17. Hydrometer. Hydrometers like these can be used to test the condition of battery acid, one cell at a time, a process that is time consuming and boring. They measure the specific gravity, which is determined by the concentration of sulfuric acid, as explained in the text. Be careful when you do this so you don't splash sulfuric acid on yourself.

to the concentration of battery acid—the higher the sulfuric acid concentration, the higher the specific gravity. If significant differences in the specific gravity of the battery acid are detected in the cells of a battery bank, it is time to equalize.

Checking voltage of the batteries and the specific gravity of the cells may involve more work than you'd like and may not be necessary if you pay attention to weather and battery voltage or adhere to a weather-sensitive periodic equalization regime. Small-wind-system expert Mick Sagrillo recommends using a wind turbine to equalize batteries in off-grid systems, if possible. You'd be amazed at how well it works, as was solar electric expert Johnny Weiss, founder of Solar Energy International in sunny Colorado. Johnny was astonished when he observed a newly installed wind turbine charge a battery bank in a PV/wind hybrid system. The wind turbine, installed during a workshop taught by Mick, was added because the client wanted to wean herself from a gasoline-powered gen-set. After living with the wind/PV hybrid system for a year, the owner sold the gen-set, as she no longer needed it. That's a properly sized PV/wind hybrid system! Unfortunately, this solution won't work for everyone. Conditions have to be right for this to work—that is, you need to have good solar *and* wind resources, and may need to be willing to curtail usage when both resources are scarce.

As a final note on the topic of equalization, be sure only to equalize flooded lead-acid batteries. Sealed batteries generally cannot be equalized! If you do, you'll very likely ruin them. (Check with the battery manufacturer, as there are some ways to carefully equalize certain sealed batteries.) Also, don't equalize batteries fitted with Hydrocaps, discussed next. Hydrocaps are installed to reduce battery watering. They must be removed prior to equalization or they will overheat and melt.

Reducing Battery Maintenance

Battery maintenance could take 10 to 30 minutes to an hour each month. (It takes about 10 minutes to check the cells in a dozen batteries but may take a half hour to check them and add distilled water to each cell if battery fluid levels are low.) If the batteries need a lot of distilled water, your time commitment will be greater. If you have a larger battery bank, say with 24 batteries, you could be in the battery room for an hour or more.

If this sounds like too much work, you'll be delighted to know that there are ways to reduce battery maintenance. One way is to install sealed batteries, although they're best suited for grid-connected systems with battery backup.

Another way to reduce battery maintenance is to replace factory battery caps with Hydrocaps, shown in Figure 7.18. Hydrocaps capture much of the hydrogen and oxygen gases released by batteries when charging under normal operation. Hydrocaps contain a small chamber filled with tiny beads containing a platinum catalyst. Hydrogen and oxygen escaping from the cells react in the presence of this catalyst to form water, which then drips back into the batteries. Because of this, Hydrocaps reduce water losses by 50% to 90%, according to several sources, greatly reducing watering requirements.

Another option is the Water Miser cap. Water Miser caps are molded plastic "flip-top" vent caps. Plastic pellets inside these caps capture most of the moisture and acid mist escaping from the battery fluid, reducing sulfuric acid fumes in the battery room and preventing terminal corrosion. However, they don't capture hydrogen and oxygen. While battery water loss is reduced, and less frequent watering is required, Water Miser caps only reduce water loss about 30%. They can, however, be left in place during equalization.

Of the two, I prefer Hydrocaps. They cost a bit more and must be removed before batteries are equalized (so don't throw out the factory-installed caps; you'll need to put these on during equalization), but they reduce water losses much more than their competitor, which reduces the amount of filling you'll need to do.

Yet another—even better—way to reduce maintenance is to install a battery-filling system, shown in Figure 7.19. Battery-filling systems consist of a series of plastic tubes connected to specially made battery caps (on each cell). Each cap is fitted with float valve (float valves function a bit like the valves in snorkels). In automatic systems, the plastic tubing is connected to a central reservoir containing distilled water. Valves in the battery caps open when the electrolyte level in a cell drops. Distilled water flows by gravity from the central reservoir. When the cell is full, the float valve stops the flow of water. (Remember to keep the distilled water reservoir filled at all times!)

In manually operated systems, the plastic tubing on each string of batteries is connected to a plastic tube fitted with a hand-operated pump. One end is immersed in a container of distilled water, for example, a one-gallon jug of distilled water. A quick

FIGURE 7.18. Hydrocaps. Hydrocaps shown here are special battery caps that convert hydrogen and oxygen released by batteries into water, which drips back into the cells of flooded lead-acid batteries, reducing watering. Credit: Joe Schwartz.

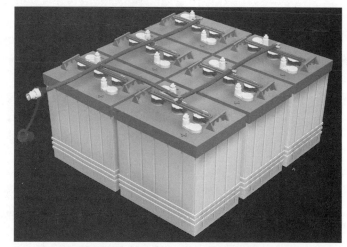

FIGURE 7.19. Battery-filling System. The Pro-Fill battery-filling system greatly reduces the time required to fill batteries and makes the chore much easier. Credit: Jan Watercraft.

FIGURE 7.20. Battery Filler Bottle. Battery filler bottles work well for systems in which batteries are accessible. Credit: Dan Chiras.

coupler allows the tubing to be connected to the pump and the distilled water supply. Water flows into each cell to the proper level. Pressure buildup tells the operator when all the cells have been filled to the proper level. For a 12-battery battery bank, the entire process takes less than five minutes.

I used a manually operated battery watering system for many years with great success. It is manufactured by Flow-Rite in Michigan. The product is called Pro-Fill and is available online. This battery filling system converted battery maintenance from a lengthy, boring, onerous task to a pleasure. What used to take me a half hour or more became a five-minute job. Instead of dreading it, and putting off, I eagerly watered my batteries each and every month.

Although battery-filling systems work well, they're costly. A system for 12 batteries in a 24-volt system might cost you as much as a battery. Although a bit pricey, such systems quickly pay for themselves in reduced maintenance time and ease of operation. The convenience of quick battery watering could overcome the procrastination that often leads to damaged batteries, a mistake that can be very costly.

A cheaper alternative, though more time consuming, is a half-gallon battery filler bottle (Figure 7.20). It comes equipped with a spring-loaded spout (similar to the float valves in battery-filling systems). This device allows you to fill each cell—one at a time and very precisely—because the valve shuts off the flow of distilled water when the battery fluid level comes to within an inch of the top of the cell. You can order them from a battery supplier, automobile parts store, or online, for $10 to $20.

Charge Controllers

Now that you understand how batteries work and, perhaps even more important, how to take care of them, let's turn our attention to the charge controller, a device that helps us care for our batteries.

A charge controller is a key component of most battery-based PV systems. If you're installing an off-grid system or grid-connected system with battery backup, you'll need one. A charge controller performs several functions, the most important of which is preventing batteries from overcharging (Figure 7.21).

How Does a Charge Controller Prevent Overcharging?

To prevent batteries from overcharging, a charge controller monitors battery voltage at all times. When the voltage reaches a certain predetermined level, known as the *voltage regulation* (VR) *set point*, the controller either slows down or terminates the flow of electricity into the battery bank (the charging current), depending on the design.

Charge controllers come in several basic designs. Let's begin with the earliest types of charge controllers: *shunt charge controllers* and *series charge controllers*. Although they are now considered dinosaurs, you may still find them in really old battery-based systems or extremely cheap on the internet. Stay away from them.

As shown in Figure 7.22, a shunt charge controller contains either an on/off switch or a variable resistor. When a charge controller such as this is operating, electricity flows out of the PV array through the charge controller then to the battery bank. When the battery voltage reaches the voltage regulation set point, the shunt is activated. Current flows through the shunt and then back to the array. No more current is delivered to the battery bank. In a switch-type charge controller, the switch closes, completely interrupting the flow of current to the batteries. In a resistor-type charge controller, the resistance to current is high when the charge controller is feeding the battery. As the battery voltage rises, however, the resistance begins to decrease, gradually reducing the flow of current. When the batteries reach the voltage regulation set point, the resistor offers no resistance to electrical current. As a result, all electrical current bypasses (is shunted) the batteries.

FIGURE 7.21. Charge Controller. Charge controllers like the one top left from Apollo Solar regulate the flow of electricity to the batteries in off-grid and grid-connected systems with battery backup. Some charge controllers contain maximum power point tracking circuitry to optimize array output and other features as well, like digital meters that display data on volts, amps, and electricity stored in battery banks. Credit: Apollo Solar.

Both types of shunt controllers close the circuit or short out the array, which protects the batteries from overcharging. It won't harm the PV modules, however, because they are designed to handle current. (Remember, they are current limited.) How long does the shunt remain operational?

Current flows through the shunt until the battery voltage drops to a predetermined setting, known as the *array reconnect voltage set point*. In a charge controller equipped with an on-off switch, the switch opens entirely, allowing current to flow

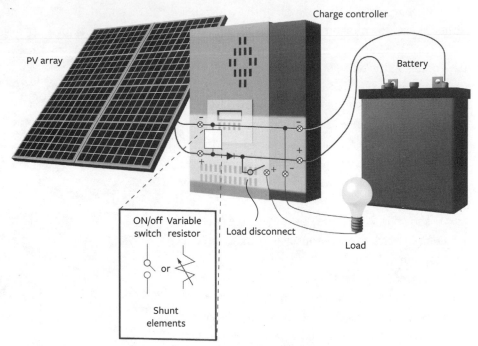

FIGURE 7.22. Shunt Charge Controller. Shunt charge controllers short circuit the array through an on-off switch or a variable resister. Credit: Anil Rao.

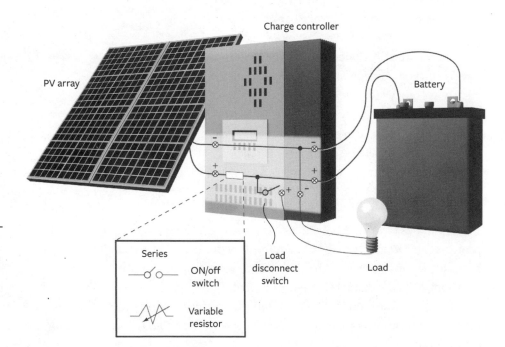

FIGURE 7.23. Series Charge Controller. Series charge controllers disconnect the array, that is, create an open circuit through a series element, either an on-off switch or a variable resistor, as shown here. Credit: Anil Rao.

back to the batteries. In a charge controller equipped with a variable resistor, resistance gradually increases, sending more current to the batteries.

The second type of charge controller is a series charge controller. As shown in Figure 7.23, some series charge controllers contain an on/off switch wired in series. When open, it stops the flow of current to the battery bank.

The series element in a series charge controller can also be a variable resistor that gradually increases resistance as battery voltage climbs toward the voltage regulation set point. This reduces the flow of current to the batteries.

Shunt and series charge controllers were replaced by devices that prevent overcharging in an entirely different manner, known as *pulse width modulation*, or *PWM*.

PWM charge controllers regulate the flow of DC electricity to a battery bank by feeding them pulses of electricity of different duration. The lower the battery charge, the longer the pulses. The higher the battery state of charge, the shorter the pulses. In these charge controllers, the length of the pulse determines how much current flows into a battery. As the batteries reach their full charge, the duration of pulses (length of each burst) is reduced. Batteries receive less electricity.

PWM is controlled by a computer that controls a switch between the PV array and the battery.

PWM charge controllers have themselves been largely replaced by a fourth type, the MPPT charge controller. As you may recall from Chapter 3, MPPT stands for *maximum power point tracking*. Like other types of charge controllers, an MPPT charge controller adjusts charging rates based on the voltage of the battery bank, and hence the level of level of charge (how full the batteries are).

Although they perform the same function, these technologies are quite different. By and large, most battery-based systems use MPPT charge controllers. Unlike PWM charge controllers, MPPT charge controllers permit array voltages much higher than battery voltages. When using a PWM charge controller, the array voltage must match the battery voltage. In addition, MPPT charge controllers work well at a wide range of temperatures, especially cold temperatures. PWM charge controllers operate best in warm temperatures.

PWM charge controllers are recommended for use in smaller PV systems. MPPT charge controllers are suited for larger arrays. Charge controllers can also be programmed to divert power to auxiliary loads when batteries are full. This feature is used in hybrid battery-based systems that couple PV with wind or microhydro.

As shown in Figure 7.24, when batteries in a hybrid system reach the voltage regulation set point, the charge controller sends surplus current to a diversion load. A diversion load is an auxiliary load, that is, a load that's not critical to the function

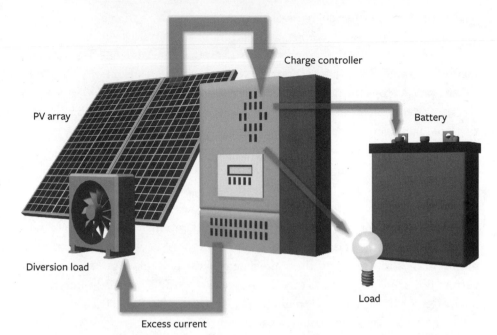

FIGURE 7.24. Diversion Charge Controller. Charge controllers can be programmed to divert surplus electricity to ancillary (nonessential) loads when batteries are full, allowing a homeowner to get more out of his or her PV system. Credit: Anil Rao.

Preventing Reverse Current Flow

Solar modules conduct electricity. When their voltage is higher than the voltage of a battery bank, for example, when the Sun is shining, electricity flows from the modules to the battery bank. At night, however, when the PV array is no longer producing electricity, current can flow from the batteries back through the array. That's because the voltage of the battery bank is higher than the voltage of the modules.

To prevent this reverse current flow, charge controllers are equipped with a diode in the circuit. Diodes allow electricity to flow in only one direction. This diode is installed in such a way that electricity can flow from the array to the module. It cannot flow in reverse.

Were it not prevented, reverse current flow will slowly discharge a battery bank. In most PV systems, battery discharge through the modules is fairly small, and power loss is insignificant. However, reverse current flow is much more significant in larger PV systems. Fortunately, all charge controllers installed in homes deal with this problem automatically.

of the home or business. In wind-electric systems, it is usually a heating element. Heating elements may be placed inside water heaters or may be in wall-mounted resistive heaters that provide space heat. In windy locations like Wyoming, they can provide quite a lot of additional heat.

In hybrid systems, excess power may be available during the summer months as well. In these instances, the diversion load may consist of irrigation pumps or fans that exhaust hot air from barns with livestock.

Why Is Overcharge Protection So Important?

As you now know, charge controllers regulate the flow of electricity to a battery bank in a controlled manner. They also prevent batteries from overcharging.

Overcharge protection is important for flooded lead-acid batteries and sealed batteries. Without a charge controller, the current from a PV array flows into a battery in direct proportion to irradiance, the amount of sunlight striking it. Although there's nothing wrong with that, problems arise when the battery reaches full charge. Without a charge controller, excessive amounts of current could flow into the battery, causing battery voltage to climb to extremely high levels. High voltage over an extended period causes severe outgassing, which causes batteries to boil, which can lead to the loss of water and sulfuric acid. Water loss can expose the lead plates to air, resulting in their demise. Overcharging can also result in internal heat production. Overheating can cause the lead plates to corrode, decreasing the cell capacity of the battery, which leads to premature death.

Some overcharge is tolerated by a flooded lead-acid battery, so long as the fluid levels don't drop below the top of the lead plates and the electrolyte is replenished. However, overcharging is extremely harmful to sealed batteries. As noted earlier, it can result in a pressure buildup inside a sealed battery that causes water and electrolyte to escape through the pressure-release valve—and there's no way for the water to be replaced.

Overdischarge Protection

Charge controllers protect batteries from high voltage, but also often incorporate overdischarge protection—circuitry that prevents the batteries from deep discharging. When the weather's cold, overdischarge protection also protects batteries from freezing. This feature is known as a *low-voltage disconnect*.

Charge controllers prevent overdischarge by disconnecting loads—active circuits in a home or business. Figure 7.23 shows the disconnect switch.

Overdischarge protection is activated when a battery bank reaches a certain preset voltage or state of charge. The low-voltage disconnect not only protects batteries, it protects loads, some of which may not function properly, or may not function at all, at lower-than-normal voltages.

What Else Do Charge Controllers Do?

Although the main purpose of a charge controller is to prevent overcharge and overdischarge of batteries, charge controllers often perform a number of additional functions. They may, for instance, control loads—that is, switch loads on and off, depending on the time of day. This feature is commonly used to control PV-powered outdoor DC lighting, for example, to illuminate signs or parking lots. In these systems, the solar array essentially becomes a photo sensor. When the current or voltage from the array drops at the end of the day, the charge controller automatically switches on lights.

Some charge controllers may also be wired to outside sensors, for example, temperature or water-level sensors. Temperature sensors allow automatic control of cooling loads, for example, evaporative coolers, and water-level sensors control irrigation pumps.

Finally, charge controllers may control automatic or user-activated equalizations. The charge controller in my off-grid system in Colorado, for instance, has a switch that can be turned on to equalize the batteries from the PV array.

Additional Considerations

When shopping for charge controllers, you will note that they are rated in amps. Solar charge controllers are rated at either 40, 60, or 80 amps. The amp rating refers to the amount of current a controller can handle from an array. Thus, arrays and controllers must be carefully matched. You wouldn't want to install a 40-amp controller in a PV system with an array that could produce 50 amps under peak sun conditions. Exceeding the amperage ratings on a controller can overload internal circuits, destroying the unit.

The charge controller-rated current (the amps listed on the equipment) must be at least 125% of the maximum output of the PV array. (I discussed how this can be calculated in Chapter 3.) The 25% safety margin permits a controller to survive cooling that occurs when clouds temporarily block the Sun. When the clouds move on, sunlight striking the array results in a spike in amperage, albeit a transitory increase. The 25% margin of safety also helps protect the equipment from occasional

1250-watts/m² days. If you plan to expand your PV system in the future, you may want to consider buying a larger charge controller.

Generators

Another key component of off-grid systems is the generator (Figure 7.25). Generators (also referred to as "gen-sets") are used to charge batteries during periods of low insolation. They are also used to equalize batteries. In addition, they can be used to provide power when extraordinary loads are used—for example, welding equipment—that exceed the output of the inverter. And, lest we forget, gen-sets may be used to provide backup power if the inverter or some other vital component breaks down.

Generators are typically "hard wired" to the battery charger inside the inverter. When the generator is started, either manually or automatically, the inverter "waits" about 60 seconds for the generator to warm up and stabilize its AC output. Then it transfers all of the home loads to the generator (a transfer switch is located internally within the inverter). The battery charger then performs a "smart," three-stage charge of the system's battery. It's a good idea, before turning the generator *off*, to terminate battery charging and let the generator run for a few minutes, so it has some idle run time to cool off before being switched off.

Gen-sets for off-grid homes are usually rather small, around 4,000 to 7,000 watts. Generators smaller than this are generally not adequate for battery charging.

FIGURE 7.25. Gen-set. Portable gen-sets like these commonly run on gasoline. Credit: Cummins Power Generation.

What Are Your Options?

Generators can be powered by gasoline, diesel, propane, or natural gas. By far the most common gen-sets used in off-grid systems are gasoline-powered. They're widely available and inexpensive. Gas-powered generators consist of a small gas engine that drives the generator. Like all generators, they produce AC electricity. The electricity travels to the inverter containing a battery charger. The battery charger contains a step-down transformer that reduces the 120- or 240-volt output of the generator to a voltage slightly higher than the nominal battery voltage, which could be 12, 24, or 48 volts. The low-voltage AC electricity is then converted to DC electricity. DC electricity is then fed into the batteries.

Gas-powered generators operate at 3,600 rpm and, as a result, tend to wear out pretty quickly. Although the lifespan depends on the amount of use, don't expect more than five years from a heavily used gas-powered gen-set. And you may need to make a costly repair from time to time.

Because they operate at such high rpms, gas-powered gen-sets are rather noisy. If you are interested in quietude, Honda makes some models that are remarkably quiet (they contain excellent mufflers). If you have neighbors, you'll very likely need to build a sound-muting generator shed to reduce noise, even if you install a quiet model. Don't add a muffler to a conventional gas-powered generator. If an engine is not designed for one, adding one could damage it (it creates back pressure on the cylinder).

If you're looking for very quiet and much more efficient generator, consider one with a natural gas or propane-powered engine. Large-sized units—around 10,000 watts or higher—operate at 1,800 rpm and, therefore, are much quieter than their less expensive gas-powered counterparts. Lower speed translates into less vibration and wear and tear. That translates into longer lifespan and less noise.

Natural gas and propane are also cleaner-burning fuels than gasoline. Unlike gas-powered generators, natural gas and propane generators require no fuel handling by you. If yours is fed by natural gas, you never have to worry about running out of fuel, so long as the natural gas service remains intact. If your generator is powered by propane, you will need to keep track of propane levels in your tank.

Natural gas and propane generators are great, but they're more expensive. You could end up paying two and a half to three times more for a natural gas or propane generator than for a well-made gasoline-powered unit. For years, natural gas and propane-powered generators were available only in fairly large sizes. Today, manufacturers are producing a number of lower-wattage generators that could work well in off-grid homes.

Another efficient and reliable option to consider is a diesel generator. Diesel engines tend to be much more rugged than gas-powered engines. As a result, these heavy-duty, long-lasting machines tend to operate without problems and for long periods. Less maintenance means lower overall operating costs and less hassle. Diesel generators are also more efficient than gas-powered generators. Another advantage of diesel engines is that they can be operated on biodiesel, a fuel made from vegetable oil. (It is chemically modified vegetable oil, not straight vegetable oil.) Biodiesel is typically mixed with petroleum diesel fuel in a ratio of 80% petroleum diesel to 20% biodiesel during cold weather. In warmer weather, the mixture can be as high as 99% biodiesel and 1% conventional diesel.

Biodiesel burns cleaner than traditional diesel and is at least partly renewable because it's made from renewable resources, typically oils extracted from soybeans or canola.

Although diesel generators offer many advantages over gas-powered generators, they cost more than their gas-powered cousins. And, of course, you will have to fill the tank from time to time. They're also not as clean burning as natural gas or propane gen-sets.

Controlling Noise

William Kemp, author of *The Renewable Energy Handbook for Homeowners*, notes, "Aside from having to run a generator in the first place, the second most annoying feature of a gen-set is having to listen to it running." He goes on to say, "The best way to eliminate this problem is not to operate them at all." Although that's a great idea, it's difficult to practice. In most cases, the best you can do is to minimize run time. Do this by selecting the sunniest location on your property for your PV array. Avoid shading. A tracking array can also help increase an array's output to keep batteries charged. You can also slightly oversize the array to produce more energy than you expect you'll need. Installing a wind turbine to provide additional power, even to equalize the batteries, can also help reduce generator run time. But, whatever you do, don't skimp on your generator run time to avoid hassle or eliminate noise. If you do, you could end up sending your battery bank to a premature death.

The next best alternative to not running a gen-set or reducing run time, Kemp goes on to say, "is to locate the unit a reasonable distance from the house and enclose it in a noise-reducing shed."

To dampen noise, insulate the walls and ceilings. However, be sure to create an opening to remove exhaust and provide fresh air. (*Do not* operate a generator inside an attached garage. Dangerous fumes can enter the home.) I built a shed from leftover 2 × 6s and insulated the walls with a mixture of straw and clay. The ceiling was insulated with a mixture of wood chips and clay. I then applied earthen plaster to the inside and finished the exterior with a coat of lime-sand plaster. The floor of my shed was made from soil-cement, a mixture of the local subsoil and 5% cement. I installed a louvered fan near the exhaust outlet of the gen-set. The exhaust fan is run off one of the 120-volt AC legs (outlets) on the generator. I created an air intake on the opposite side. This shed was extremely effective in blocking noises and, because of the fan, it stayed cool inside. Never had another complaint from my neighbors!

Kemp recommends building a shed with a floating floor—that is, a floor that does not contact the walls. This keeps engine vibration from radiating outside the building. "The walls should be packed solidly with rock wool, fiberglass or, best of all, cellulose insulation. The insulation should then be covered with plywood or other finishing material, further deadening sound."

Installing a gen-set away from your home may require long-distance overhead or underground feed cable. The cable will need to be sized correctly to reduce line loss. The cable will either need to be rated for burial or installed in conduit. Be sure not to install a gen-set close to a neighbor's house to avoid noise in your own.

A generator with an automatic start switch will make your life a lot easier. This fairly inexpensive feature can be wired to the inverter (for truly automatic function) or to a manual switch inside the home. This feature will save you trips to the generator shed on the freezing cold winter days when the generator is most frequently called into service. Be sure to run wire for the automatic start circuitry when you run the power cable.

As noted earlier, yet another way to reduce noise is to install a quieter gen-set, for example, a gasoline-powered generator with a good muffler, or a lower-rpm diesel, propane, or natural gas generator. If there are no neighbors to bother, it may not be necessary to install a quiet generator or house a gen-set in an insulated shed. However, be sure the generator is protected from weather.

Other Features to Look For

Gasoline-powered generators typically come with four 120-volt AC outlets into which extension cords can be plugged. They also often contain 120-volt 20 or 30 amp outlets that are used to power battery chargers.

Be sure your generator comes with a low-oil shutoff. This prevents the generator from starting if the oil level is low or shuts it off if oil levels fall below a critical level during operation.

Another feature to consider when shopping for a generator is a run-time meter. It will cost a bit more, but will help you keep track of run time so you know when to perform scheduled maintenance.

Living with Batteries and Generators

Batteries work hard for those of those of us who live off grid. As you have seen, flooded lead-acid batteries need to be kept in a warm place, but not too warm. Their enclosures need to be vented, too. These batteries also need to be periodically filled with distilled water. And, you need to monitor their state of charge and either charge them periodically with a backup supply of power when they're being overworked or back off on electrical use. Either way, don't allow batteries to sit in a state of deep discharge for more than a few days if you can help it.

Generators in off-grid systems need a bit of attention, too. If you install one in your system, you will need to periodically change oil and air filters. If you install a manually operated generator, you'll need to fire it up from time to time to raise the charge level on your batteries or to equalize the batteries.

It is also a good idea to run a generator from time to time during long periods of inactivity, for example, over the summer when a generator is typically not used. Gasoline goes bad sitting in a gas tank. It turns into a shellac-like material that can gum up hoses and carburetors. To prevent this, add a fuel stabilizer to the tank at the end of each summer and run the motor from time to time. Finally, gasoline-powered generators can be difficult to start on cold winter days, so be sure to use the proper weight oil during the winter.

Gas-powered generators, while inexpensive, tend to require the most maintenance and have the shortest lifetime. Be prepared to haul your gas-powered gen-set in for an occasional repair. My backup generator visited the repair shop twice in 13 years for costly repairs—and I only ran it 10 to 20 hours a year!

In grid-connected systems with battery backup, you'll have much less to worry about. If you install sealed batteries, for example, you'll never need to check the fluid levels or fill batteries. Automatic controls keep the batteries fully charged. If you install flooded lead-acid batteries, be sure to check them every couple months. Write reminders on your calendar!

Batteries may seem complicated and difficult to get along with, but if you understand the rules of the road, you can live peacefully with these gentle giants and get many years of faithful service. Break the rules, and it's a sure thing you'll pay for your inattention and carelessness.

Mounting a PV Array for Maximum Output

For a solar electric system to produce the most electricity, it must be properly installed. Break any of the rules, and you'll pay a penalty in lower production for the entire lifespan of your system.

To begin, for best results, the array should be oriented so that it points directly to true south in the Northern Hemisphere or true north in the Southern Hemisphere. If the array is fixed, that is, if its tilt angle can't be adjusted, it should be mounted at an angle that provides the maximum output during the year—that's usually latitude minus five degrees. Or, the array can be installed so that the tilt angle can be manually or automatically adjusted to optimize output. (Tilt angle is most often manually adjusted, if it is adjusted at all.)

Next, the array must be mounted in a location that receives as much sunlight as possible throughout the year. As noted in Chapter 4, installers can use a Solar Pathfinder or a cell phone app to select the sunniest—least shaded—locations for arrays. If your installer doesn't use one of these (or similar) tools, and you have doubts about the result of shading or less-than-optimum orientation, consult with another installer. I'm always amazed when I show up to inspect solar electric installations to find that they are not oriented to true south or, in some cases, are heavily shaded. Figure 8.1 shows an array on a steep east-west roof, which will reduce the array's output by at least 40% to 50%.

Although the task of siting a solar array in a sunny location and orienting the array just right so that it generates the most electricity possible may seem pretty straightforward, it can be quite challenging. For example, it is not always possible to find a location, especially on existing buildings, that provides full access to the Sun all year long. This is especially difficult in cities and suburbs. Shade trees,

FIGURE 8.1. East-West Oriented Array. Irresponsible and uninformed solar installers sometimes install PV arrays on east- and west-facing roofs (like the one shown here) under the misguided notion that this orientation will only reduce PV output by 15%. That's not true for steep roofs like this one. This array, which faces due east, is in the shade shortly after solar noon each day. You can't see it, but there are also several large shade trees east of the array that block sun part of the year. And, making matters worse, the east-west oriented roof of this house shades the lower part of the array on winter days when the Sun is low in the sky. Don't make this egregious mistake. Credit: Dan Chiras.

neighboring buildings, and even portions of the building on which an array is mounted can shade it part of the time. While every effort should be made to install a PV array in a shade-free location, compromises are sometimes necessary.

Siting an array may also be influenced by aesthetic considerations. Building departments, historic preservation districts, homeowners associations, spouses, and neighbors may influence the decision, putting pressure on you to place an array in a location that could reduce its output. I had a client in Chicago whose building department insisted that their array be mounted on a north-sloping roof, so it wouldn't be so visible from the street in a historic district. (We worked with the building department to help them understand why this location wouldn't work, but we lost the battle, so the client decided to sell the property and build elsewhere!)

This chapter covers high-performance installation—that is, installation that will ensure a safe, productive PV system. It will introduce you to a number of installation options that provide some flexibility in meeting the twin goals of maximum output and aesthetics.

What Are Your Options?

Mounting options for the PV array are usually dictated by shading issues and aesthetics. Generally speaking there are two options: building-mounts or ground-mounts. Building-mounted arrays are most common in densely packed areas, notably cities and suburbs. Most building-mounted arrays in these areas end up on roofs so the arrays are above trees and buildings. In these locations, modules are generally mounted on lightweight aluminum racks. That said, building-mounted arrays also include *building-integrated PV*, or BIPV, wherein the array is part of the roof or installed as an awning to provide shade to windows in the summer. BIPV was discussed in Chapter 3.

Ground-mounted arrays are most common in rural areas, where there's generally more real estate. Two options are available when mounting an array on the ground: rack mount and pole mount. Rack mounts are the most common by far.

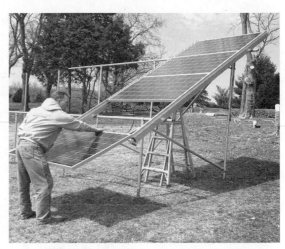

FIGURE 8.2. Several Types of Rails. Rails for mounting PV modules consist of a series of channels used to attach them to modules and mounting hardware, typically L-feet. (*left*) Most rails on the market look like this one from Renusol, and most modules are attached by midclamps and endclamps, also shown here. (*center*) PV Racking makes a rail that I use a lot. As you can see, it has slots into which installers slide the modules. No WEEBs or ground lugs and ground wire are required. (*right*) A single worker can install the modules by himself. Credit: Dan Chiras.

In this chapter, we'll discuss the various options within these two broad categories. I'll discuss each type and their pluses and minuses.

Ground-mounted Racks

My favorite location to mount PV arrays is on the ground, usually on a rack. One reason for this is that the arrays stay cooler in the summer than roof-mounted arrays. Because of this, they perform better—often substantially so. (Remember, heat decreases the output of many PV modules.)

In a ground-mounted PV array, modules are attached directly to extruded aluminum rails like those featured in Figure 8.2. The rails are attached to a steel substructure (Figure 8.3). Together the rails and the steel substructure form the rack.

Racks must be secured to the ground. That's accomplished by a steel-reinforced concrete foundation, like

FIGURE 8.3. Steel Rack for Ground-mounted Arrays. Ground-mounted racks typically include a steel rack that supports the aluminum rails like this one the author installed at Owensville Middle School in Owensville, Missouri, with the help of several hard-working students. That's my business partner, Tom Bruns, who works with me on all my installations. Credit: Dan Chiras.

FIGURE 8.4. Steel Augers. Galvanized steel augers like the one shown here can be used to anchor a PV array to the ground. Credit: Dan Chiras.

the one shown in Figure 8.3, to prevent the array from taking flight in strong winds. This adds cost to an installation, but the benefits of this approach often outweigh the additional cost. In parking lots, PV arrays can be anchored to large above-ground concrete blocks. (You'll see a photo of this in Figure 8.20.)

Some companies attach steel pipes in the ground by drilling galvanized steel augers into the soil using a skid steer equipped with an auger head. As shown in Figure 8.4, steel augers are very large screws. They are drilled into the ground and left in place. The steel rack is bolted to the top of the auger. The threads (flights) hold them in the ground.

Once made from steel, rails for PV system are now manufactured from extruded aluminum. As shown in Figure 8.2, all rails have channels that run their full length. The channels allow installers to attach modules via stainless steel bolts. The channels also allow installers to secure the rails to the steel substructure (Figure 8.6).

One innovative rail that I use to cut down on installation time and frustration on ground-mounted PV systems is manufactured by PV Racking in Pennsylvania. As shown in Figure 8.2b, it consists of two channels into which the installers slide the modules. This feature greatly reduces installation time and cost.

In ground-mounted arrays, rails may attach directly to the steel substructure, as shown in Figure 8.6. Some installers secure adjustable legs to the rails, then attach the legs to the steel substructure. This allows the homeowner to adjust the tilt angle of the array during the year to improve performance.

Ground-mounted racks come in two basic varieties: fixed and adjustable. Fixed racks are typically set at the optimum angle (the latitude of a location minus 5 degrees. If you live at 45° north latitude, for instance, the array should be set so the tilt angle is 40°).

FIGURE 8.5. Steel Augers Being Installed. We use this machine or a skid steer equipped with an auger attachment to install augers to mount PV systems. One thing I like about this system is that it minimizes ground disturbance and makes for a much neater installation. Credit: Dan Chiras.

Adjustable racks, as their name indicates, can be adjusted to maximize annual energy production by an array. Most arrays are adjusted seasonally, usually twice a year. Adjustability is typically provided by telescoping back legs. Changing the length of the back legs allows the tilt angle to be increased or decreased (Figure 8.7). Some telescoping legs come with pre-drilled holes or slots that correspond to different tilt angles. This limits your options. Most legs I've used on ground and roof mounts, however, are infinitely adjustable. A telescoping leg consists of two pieces of aluminum. The smaller piece (C-channel) slides inside the larger-piece (square tubular aluminum) illustrated in Figure 8.7. A single bolt secures the legs at any point you want.

Ground-mounted Pole-mounts: Fixed and Tracking Arrays

Although not as widely used as ground-mounted racks, pole-mounted racks may be just what the doctor ordered for your location—if you are installing a small array (around 3 kW or smaller). As shown in Figure 8.8, in this scenario, the array is attached to an aluminum rack that's mounted on a sturdy steel pole anchored securely in the ground. The poles are typically embedded in a steel-reinforced concrete base.

PV arrays are usually mounted on the top of the pole, although they can also be mounted on the side of poles. Top-of-pole mounts allow operators to seasonally adjust the tilt angle of the array to accommodate changes in the altitude angle of the Sun.

The steel pipe that supports a pole-mounted array can range from 2½-inch schedule 40 steel pipe for small arrays up to 8-inch schedule 40, or schedule 80 for larger arrays. (The term "schedule" refers to the thickness of the pipe wall and hence the strength of the pipe.) Square steel tube is also used for larger

FIGURE 8.6. Close-up of Rail Attached to Steel Substructure. This is my favorite rail on the market; it's made by PV Racking in Pennsylvania. Here you can see the bracket that attaches the rail to the underlying steel tubing in this ground-mounted system at a customer's house. Credit: Dan Chiras.

FIGURE 8.7. Telescoping Legs. Telescoping legs, like the ones on TEI's first solar array in 2009, allow for infinite adjustability within a range dictated by the length of the two aluminum pieces, as shown here. Here, students and a veteran installer who helped teach this workshop check the angle and distance from the front to the back rail. Credit: Dan Chiras.

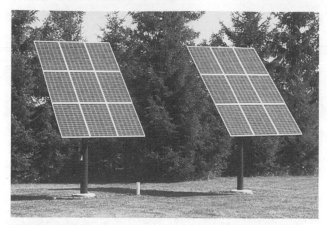

FIGURE 8.8. Pole-mounted Array. Each array is mounted on a pole anchored to the ground and is installed at the optimum angle for its location. Pole-mounted arrays provide some flexibility in placing an array on a lot. Credit: Anthony Powell.

arrays. Square tube won't turn in a concrete base when the winds blow on an array and is, therefore, better than round pipe.

For round pipe, pipe size, which is measured in inches, refers to the inside diameter (ID) of the pipe. The ID of a 6-inch schedule 40 pipe is 6 inches. Its outside diameter (OD) is 6⅝ inches. Rack manufacturers specify the ID diameter and the schedule of pipe needed to mount its rails. Because steel pipe is expensive to ship, installers typically purchase pipe locally. It is available from steel or plumbing suppliers, which are common even in rural areas. To be sure that an array is firmly anchored in the ground, the steel-reinforced concrete base must be properly engineered. The concrete must also be poured a couple of weeks *before* the array is mounted on the pole so it cures properly.

Pole-mounted arrays fall into two categories: fixed or adjustable. Fixed arrays are immovable. Modules are mounted so the azimuth and tilt angles remain constant throughout the year at an angle that maximizes output.

Adjustable pole-mounted arrays fall into two categories: manually or automatically adjustable. A manually adjusted array allows the owner to change the tilt angle of the array by season to increase the array's output. Although a homeowner can adjust an array four times a year—spring, summer, fall, and winter—most people make the simple adjustments twice a year. The first adjustment is made on or around the spring equinox (the first day of spring). At that time, the array is adjusted at an angle equal to the latitude of the site minus 15 degrees. Reducing the tilt angle positions the array so it captures more of the high-angled summer sun. The second adjustment is made on or around the fall (autumnal) equinox (the first day of fall). At that time, the array is tilted to an angle equal to the site's latitude plus 15 degrees. This adjustment positions the array so that it more directly faces the low-angled winter Sun.

Besides increasing output, tilting an array to take advantage of the low-angled winter Sun also allows the array to shed snow better than a flatter array would. Arrays at higher latitudes that are mounted at very steep angles may also receive energy from sunlight reflecting off snowy landscapes.

Seasonal adjustments of the tilt angle can increase the output of an array by 10% to nearly 40%, depending on the latitude of the site, location of the array on one's

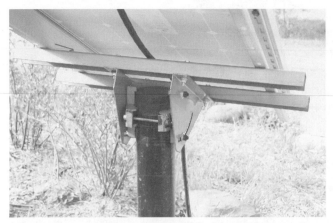

FIGURE 8.9. Adjustment for Pole-mounted Array. The tilt angle of a seasonally adjustable pole-mounted array can be changed to accommodate the changing altitude angle of the Sun to increase electrical production. Credit: Dan Chiras.

FIGURE 8.10. Magnetic Angle Finder. This inexpensive device is used to check the angle of roofs and arrays. Credit: Dan Chiras.

property, and shading. In one of my ground-mounted arrays, I estimated that the output of the array, were it to remain at a fixed angle, would be about 5,200 kWh/year. My students and I adjust it twice a year, and the array consistently produces about 6,200 kWh/year.

Adjusting a small, ground-mounted array may be as simple as loosening a nut on a bolt on the back of the array mount, then tilting the array up or down (Figure 8.9). A magnetic angle finder, shown in Figure 8.10, can be used to set the tilt angle. Once the angle is correct, the nut is tightened. Manually adjusting a large pole-mounted array, however, can be a challenge. See the sidebar, "Tilt Angle: Getting It Right" for advice on tilt angle adjustment.

Another very alluring option for a pole-mounted array is a *tracking array*. Tracking arrays are designed to track the Sun from sunrise to sunset (Figure 8.11). They are designed to orient the array so that it points directly at the Sun (or as close as possible) at all times. This ensures maximum absorption of solar radiation. Two types of trackers are available: single-axis and dual-axis.

FIGURE 8.11. Tracking Array. This tracking array was installed by Dan and fellow students at a workshop in Minnesota sponsored by the Midwest Renewable Energy Association located in Custer, Wisconsin. Credit: Dan Chiras.

A *single-axis tracker* adjusts one angle, the azimuth angle, based on the position of the Sun at different times of day. Single-axis trackers rotate the array from east to west, following the Sun from sunrise to sunset. As you may recall, PV modules generate the most electricity when directly facing the Sun (the incidence angle is 0°). While single-axis tracking arrays adjust for azimuth angle, they do not take into account changes in the tilt angle through the solar cycle. In these arrays, the tilt angle is either fixed year round or can be manually adjusted two to four times a year to account for seasonal differences in the altitude angle of the Sun.

Tilt Angle: Getting It Right

Most solar experts recommend adjusting an array twice a year: the first time at or near the spring equinox; the second time at or near the fall equinox. The rule of thumb is that during the summer the tilt angle should be set at latitude minus 15 degrees. If you live at 40° north latitude, for instance, the tilt angle should be about 25° in the summer. To maximize winter output, the tilt angle of the array should be set at the latitude plus 15 degrees. If your home or business is at 40° north latitude, then the tilt angle should be about 55° in the winter.

Although the rules of thumb work pretty well, I recommend checking the NASA tables on the website "Surface Meteorology and Solar Energy" for more precise recommendations. The NASA data is based on measurements that take into account seasonal variations in cloud cover. As an example, Table 8.1, shows a site in south central Tennessee at 35° north latitude. According to the rule of thumb, the optimum year-round tilt angle at this location should be 35°. According to the NASA site, however, the optimum angle (found on the last row of the table—far right column) for best year-round performance for a fixed array is 30°. This is due, in large part, to the fact that there is more sunlight in the summer months. Days are longer and the Sun is overhead, resulting in greater daily irradiation. A slightly "flatter" array (lower angle) will produce more electricity.

The rule of thumb also falls apart if you seasonally adjust your array according to it—raising in 15° in the winter and lowering it 15° in the summer. In our example in south central Tennessee, the best wintertime tilt angle would be 50° (35° + 15°). The best summertime angle would be 20° (35° − 15°). The NASA website shows that in the winter the optimum tilt angle is actually 54° to 58°—slightly higher than the 50° general rule of thumb. (Check out optimum angles for November, December and January). In the summer, the optimum tilt angle is 2° to 13°—much lower than the 20° rule-of-thumb recommendation.

The performance difference between the two methods of determining tilt angle is small but significant. For example, using the data from Table 8.1 for the site in Tennessee, if you were to adjust the tilt angle according NASA recommendations, you would get a 0.7% improvement over the rule-of-thumb tilt angle in the winter, and a 1.7% improvement in the summer. Even though the differences are small, over the lifetime of a system those small annual increases (especially in a large system) can result in thousands of extra kilowatt-hours of electricity.

A tracking array that moves the array so that it adjusts for changes in both the altitude angle and the azimuth angle of the Sun is known as a *dual-axis tracker*. Dual-axis trackers follow the Sun's azimuth angle from sunrise to sunset—just like single-axis trackers. They also adjust the tilt angle to follow daily and seasonal changes in the Sun's altitude angle. As you learned in Chapter 2, the altitude angle of the Sun increases from 0° at sunrise to a high point at solar noon, then decreases to 0° again at sunset. Altitude angle also changes from one day to the next. For example, the altitude angle of the Sun at solar noon (halfway between sunrise and sunset) increases each day as the Sun "moves" from its lowest point at the winter solstice to its highest point at the summer solstice. From summer solstice to the winter solstice, the altitude angle at any time of the day decreases from one day to the next.

Dual-axis trackers are most useful at higher latitudes because days (hours of sunlight) are much longer in the summer in these regions. In the tropics, where daylength remains the same throughout the year, a single-axis tracker will perform as well as a dual-axis tracker, provided it has access to the Sun from sunrise to sunset.

Although trackers increase an array's annual output, the greatest impact on energy production occurs during the summer when days are longer and typically sunnier. Output improves less during the short, often-cloudy days of winter. For those installing grid-tied systems, a tracker's excess summer production helps offset winter's lower production—but only if you have annual net metering or monthly net metering that reimburses you at retail. Monthly net metering that reimburses at wholesale (avoided cost) won't help one bit, because utilities typically only pay customers one-fourth of the electricity's value.

Table 8.1. Monthly Averaged Radiation Incident On An Equator-pointed Tilted Surface (kWh/m²/day)

Lat 35 Lon -87	Jan	Feb	Mar	Apr	May	Jun	Jul	Aug	Sep	Oct	Nov	Dec	Annual Average
SSE HRZ	2.23	2.93	4.01	4.98	5.52	5.80	5.79	5.28	4.72	3.75	2.60	2.04	4.14
K	0.44	0.45	0.48	0.49	0.49	0.50	0.51	0.51	0.53	0.53	0.48	0.44	0.49
Diffuse	0.97	1.27	1.66	2.05	2.36	2.47	2.38	2.13	1.71	1.30	1.02	0.88	1.69
Direct	3.10	3.45	4.16	4.62	4.76	4.94	5.09	4.85	5.05	4.81	3.71	3.04	4.30
Tilt 0	2.20	2.83	3.96	4.95	5.49	5.76	5.76	5.25	4.65	3.72	2.55	2.03	4.10
Tilt 20	2.84	3.39	4.42	5.13	5.41	5.56	5.60	5.32	5.08	4.47	3.28	2.70	4.44
Tilt 35	3.15	3.61	4.51	4.97	5.05	5.11	5.18	5.08	5.10	4.76	3.63	3.04	4.44
Tilt 50	3.30	3.65	4.38	4.58	4.47	4.43	4.53	4.60	4.87	4.80	3.79	3.22	4.22
Tilt 90	2.84	2.86	3.04	2.69	2.33	2.21	2.28	2.55	3.15	3.69	3.21	2.84	2.81
OPT	3.31	3.65	4.51	5.13	5.52	5.76	5.76	5.35	5.12	4.81	3.79	3.24	4.67
OPT ANG	55.00	45.00	34.00	18.00	6.00	2.00	4.00	13.00	29.00	45.00	54.00	58.00	30.10

FIGURE 8.12. Photo Sensor on Pole-mounted PV Tracker. This electronic eye sends signals to the controller which adjusts the tilt angle and azimuth angle of the array to track the Sun across the sky. Credit: Dan Chiras.

Trackers are also less useful for off-grid systems because summer surpluses are typically wasted. Once the batteries are full, the array's output has nowhere to go. Charge controllers prevent the electricity from overcharging the batteries or, in the case of diversion charge controllers, dump the surplus into nonessential loads, as explained in Chapter 7.

Trackers operate either actively or passively. Active systems rely on electric motors to adjust the array's angles. In many tracking arrays, the Sun's position is detected by photosensors mounted on the array. These sensors send signals to a controller, a small computer that controls the electric motors. They adjust the position of array angles as the Sun's altitude and azimuth angles change during the day. In some systems, tracking is controlled by a computer that's programmed with the altitude and azimuth angles of the Sun for every day of the year.

The electric motors in active tracking arrays may be powered by either (1) DC electricity directly from the solar array, (2) DC electricity from a battery bank in battery-based systems, or (3) AC electricity from the house or business. Most installers choose the last option: powering the motors with AC electricity. AC electricity is converted to DC by an AC-to-DC converter, typically mounted on the pole.

Running AC from a home requires installers to bury an electric line that runs from the main panel in the house to the array. It runs alongside the DC wires running to the inverter inside the house or business. The wires are often run in plastic conduit to protect them from being damaged by rocks in the soil. If you are considering this, remember that the National Electric Code (Article 690) prohibits installers from running these wires in the same conduit with the wires from the PV array to the inverter.

Figure 8.12 shows a photosensor on an array I helped install as part of a Midwest Renewable Energy Association workshop. Figure 8.13a shows the electric motor that adjusts the azimuth angle and Figure 8.13b shows the motor that adjusts the tilt angle.

Passive-tracking systems require no sensors or motors. However, they track only the Sun's azimuth angle—from sunrise to sunset. As shown in Figure 8.14, passive trackers are equipped with fluid-filled tubes positioned on either side of the array. They are connected by a small-diameter pipe. When sunlight strikes the tube on

FIGURE 8.13. (*a*) Motor That Adjusts Azimuth Angle. This motor sits atop the pole and adjusts the azimuth angle of the array to track the Sun on its east-to-west path across the sky. (*b*) Motor That Adjusts Tilt Angle. A smaller electric motor adjusts the tilt angle of the array to accommodate the ever-changing altitude angle of the Sun. Credit: Dan Chiras.

the left, for example, right after sunrise, it heats the liquid, causing it to vaporize and expand. Expansion, in turn, forces liquid to flow into the tube on the right. This causes the weight to shift to the right, which causes the array to shift, tracking the Sun as it "moves" across the sky.

FIGURE 8.14. Passive Tracker. Passive trackers like this one do not require motors or photo sensors. How they work is explained in the text. Credit: Zomeworks.

How Important Is Tracker Accuracy?

As most readers will know by now, PV modules generate the most electricity when directly facing the Sun. But how much output is lost if the tracker is a little off—if it doesn't point the array *directly* at the Sun? The answer is—not much. The relationship between tracking error and output is a simple mathematical function, the cosine function. Tracking error is, as the name implies, how far off an array is from perfect alignment with the Sun. (It's the angle between a line perpendicular to the face of the array and another line from the array to the sun.) As the following table shows, small tracking errors have very little effect on PV output.

Tracking Error (degrees)	Relative PV Output
0°	100%
5°	99.6%
10°	98.5%
15°	96.6%
20°	94.0%
25°	90.6%

Passive trackers systems are touted as being dependable as gravity and reliable as the Sun, but they have some limitations.

Both active and passive trackers increase array output, but active trackers are more efficient. That's because active trackers rotate back to sunrise position, facing east, at the end of each day. Passive trackers follow the Sun from sunrise to sunset but, at the end of the day, remain pointing west toward the sunset. It's not until sunrise that they'll move back into position. Here's how it works: As the morning Sun rises in the east, it heats the unshaded west-side canister. This forces liquid into the shaded east-side canister. As liquid moves to the east-side canister, the tracker slowly rotates back to the east. Unfortunately, it takes an hour or two for the passive tracker to return to position. During this time, little, if any, electricity is being produced. It can't. The array has its back to the Sun.

Because an active tracking array returns to its east-facing position immediately after sunset, it begins producing electricity an hour or two earlier each day than a passive-tracking array. That's one reason why an active tracking array generates more energy than an array with passive tracking. Another is that passive trackers can also be deflected by wind. On windy days, they may not always be pointing directly at the Sun.

For optimum performance, arrays equipped with trackers must receive dawn-to-dusk sun. "There's no point in buying a tracker if your site doesn't begin receiving sunlight until 10 in the morning, or loses it at 2 in the afternoon due to shading from hillsides, trees, or buildings," notes Ryan Mayfield in an article in *Home Power* magazine. So, which should you buy—a single—or a dual-axis tracker?

Even though a dual-axis tracker may seem as if it would produce a lot more electricity than a single-axis tracker, the benefit is actually only marginal. As shown in Figure 8.15, a single-axis tracker generally results in the greatest improvements in array output. Dual-axis trackers increase output over a single axis, but only very slightly.

Should you buy an active or a passive system? The answer to this question depends on where you live. Active trackers work well in both cold and hot climates and everything in between. Passive trackers on the other hand, do not perform as well in cold climates. They tend to be sluggish and imprecise on cold winter days. That's because they depend on solar heat to vaporize the refrigerant. If you live in a cold climate, consider an active tracker. In warmer climates, passive trackers tend to be more reliable.

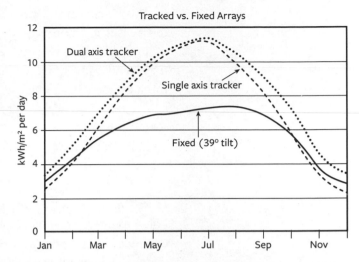

FIGURE 8.15. NREL Graph Showing Output of Tracking Arrays. This graph compares the monthly output of a fixed array, a single-axis, and a dual-axis array. A single-axis tracker dramatically increases the annual output compared to a fixed array. A dual-axis array results in very little additional gain. Credit: Anil Rao.

On the downside, active trackers require photosensors, controllers, and electric motors. The more parts, especially electronic parts, the more likely something will go wrong. Photosensors used to be the weak link in active trackers; they broke down frequently and required replacement. Fortunately, manufacturers have improved their designs, resulting in much more reliable sensors.

Today, the weakest link in active tracking arrays is the controller. The controller is a small logic board that integrates information from the photosensor that's used to regulate the motors that adjust the tilt and azimuth angles of the array. Unfortunately, this circuitry can be vulnerable to lightning. While manufacturers have improved lightning protection in active trackers, a nearby or direct lightning strike can damage a controller, necessitating replacement.

According to some folks, the best answer to the question—should I install an active or passive system?—may be neither. Although it is true that trackers increase the output of a PV array, the key question to ask is how much does this additional output cost? And, is it worth it? Or, are there other, perhaps simpler and more reliable, ways to boost the electrical output of a PV array?

Proponents of the view that "no tracker is the best tracker" point out that, in most cases, it is more economical to invest in more PV modules to boost the output of a PV array than to install a tracker array that yields the same output. A larger, fixed array will require less maintenance, too, lowering its overall cost. If you're thinking about installing a tracker, run the math, or ask your installer to run it for you, to determine whether it really makes sense. Be sure to take into account periodic controller replacement. And remember that shading at the site may make this a moot point. Trackers are only suitable for sites that have a horizon-to-horizon, open solar window.

Installing a Pole-mounted Array

Pole-mounted arrays should be installed by a professional installer or a very experienced homeowner with experience in electrical wiring and PV installation. To install a pole-mounted array, you will need to excavate a deep hole for the poured concrete foundation. Exact specifications for the excavation and size of the pole are available through manufacturers of pole-mounts. To give you an idea what you are in for: for a 120-square-foot array, the pole-mount manufacturer (Direct Power and Water) recommends using 6-inch schedule 40 pipe. For a pole that's 6 to 7 feet (1.8 to 2.1 meters) above ground, they recommend embedding 4 to 6 feet (1.2 to 1.8 meters) of pipe in a steel-reinforced concrete foundation.

If installing a system yourself, be sure to check with the local building department. Some jurisdictions require homeowners to hire a licensed engineer to specify foundation details based on local soil and wind conditions. If you're hiring a professional installer, they should provide this information and secure the permit. If you use round pipe, be sure to secure the pipe to the reinforcing steel in the foundation so it will not twist when the winds blow.

You will also need to run wire from the house or business to the array. When soil conditions permit, the wire should be buried. Include separate conduits through the foundation for the DC output and the AC power to motorized trackers. Once the pole is in place, you may need to mount a *combiner box* on it (Figure 8.16). It combines the positive and negative home-run wires from each series string in a PV array into a single positive and negative wire. It runs underground to the inverter. If the run to the house is long, the output of the combiner box can be wired to a *transition box*. (Figure 8.17). The transition box allows smaller wires from the combiner box to connect to larger-diameter wires that run to the house. Larger-diameter aluminum wire is typically used to save money and improve efficiency. It reduces voltage drop (line loss) over long distances. This wire terminates in a transition box inside or

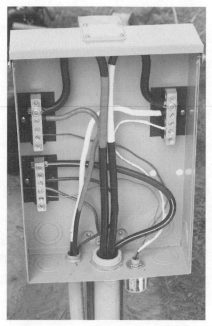

FIGURE 8.16. Combiner Box. A combiner box combines the output of the various strings in a PV array for an off-grid system. A single wire runs from the combiner box to the house. Positive wires from the PV arrays connect to circuit breakers. Two large-diameter wires run from the positive and negative bus bars (top and left, respectively) to the house. Combiner boxes for high-voltage grid-tied systems look slightly different, as they typically contain fuses rather than circuit breakers. Credit: Kurt Nelson.

FIGURE 8.17. Transition Box. In this installation for a grid-connected system in Minnesota, the array was located several hundred feet from the house, so the output of the combiner box was run to a nearby transition box, shown here. From there a single, large-diameter wire ran underground to the house. Large-diameter wire was used to reduce line loss. Credit: Dan Chiras.

outside the house. The transition box is wired into the inverter in grid-tied systems or the charge controller in a battery-based system.

Once the foundation has completely cured (don't rush it!), the array is mounted on a rack attached to the pole. The rails of racks are designed for specific modules, so be sure to specify the number and the modules you'll be using when ordering a pole-mount rack and hardware. Some installers recommend purchasing a larger rack than initially required so that more modules can be added at a later date if you decide to increase the size of the system.

If a motorized tracker is used, you'll need to mount the electric motor on top of the pole, then install a heavy piece of square steel tubing that runs perpendicular to the vertical pole (forming a T). This piece is known as the *strong back* or *torque tube*. It forms the spine of the rack. Rails and supporting cross braces are attached to the strong back (Figure 8.18). The modules are then inserted one by one into the metal rack (Figure 8.19). Modules are wired in series, creating strings that are then wired in parallel in the combiner box.

FIGURE 8.18. Pole-mount Being Assembled Showing Strong Back. Students at MREA workshop installing rails on the strong back on a pole-mount. Credit: Dan Chiras.

FIGURE 8.19. Module Being Inserted in Pole-mount. Workers carefully slide PV modules into place on this dual-axis tracker at a Midwest Renewable Energy Association workshop in Minnesota. Credit: Dan Chiras.

Installing an array on a pole-mount can be challenging, so proceed carefully. Take a workshop or two and be sure to read Joe Schwartz's two excellent articles on the subject in *Home Power* magazine (Issues 108 and 109). They're very well illustrated.

Pros and Cons of Ground-mounted Arrays

Ground-mounted arrays offer many advantages over building-mounted PV systems, but they do have a few drawbacks. One of the biggest advantages of ground-mounted arrays is they can be positioned far away from anything that might cast shade on an array. I've installed grid-tied PV arrays as far as 1,000 feet (300 meters) from homes (but had to use very large wire [4/0] to minimize voltage drop). I install most grid-tied arrays with voltages of 240 AC or DC voltages in the 350 to 450-volt range within 100 to 300 feet of the main service panel. I always compensate by using a large-gauge wire, however. To determine wire size, you or your installer must calculate the voltage loss of the proposed wire. For the run from the array to the inverter, it's best to shoot for a voltage loss of no more than 1%.

Even lower-voltage off-grid and grid-tied systems with battery backup can be located 100 to 200 feet (30 to 60 meters) from the house. In such cases, however, even larger-diameter wire will be needed to efficiently transmit the low-voltage DC.

Ground-mounted arrays placed on racks or poles can also be positioned more precisely than many roof-mounted system. That is, they can be oriented to true south and tilted at just the right angle to ensure maximum production. Ground-mounted arrays may be your only choice if a roof is unsuitable for some reason—for example, if the roof faces the wrong direction, is at an inappropriate pitch, is shaded, or is not strong enough to withstand wind loading (force of the wind that could rip an array from a roof).

Ground-mounted arrays also help to maintain cooler module temperatures than roof-mounted arrays. Most PV modules produce more electricity at cooler temperatures.

Precise positioning and cooler array temperatures increase the output of a PV array and the value of your investment in PVs. In other words, you'll get more electricity out of the system over its lifetime. This, in turn, lowers the cost per kilowatt-hour.

Ground-mounted arrays are also often a lot safer to install than roof arrays. There's no need for potentially dangerous roof work.

Another advantage is that ground-mounted arrays do not require roof penetrations like their roof-mounted cousins. Driving lag screws into a roof to install a rack and cutting holes to run wires to the balance of the system creates potential leaks in a roof. Ground mounting an array, as opposed to a roof mount, also avoids the dismantling of an array when time comes to re-roof a house. Generally speaking, PV modules will outlive most roofs, sometimes multiple re-roofings. My company just removed a PV array that had been on a roof only four years. The roof was damaged by hail. Removing then reinstalling the array cost the homeowner's insurance company $3,500.

Ground-mounted arrays generally permit easier access to solar modules than a roof-mounted array. This enables an owner to clean a dusty array, if necessary, or gently brush snow off modules to maximize output. (Though cleaning is rarely required in most regions.) Of course, ground-mounted racks are almost always easier to access than pole-mounted arrays.

Ground mounting also permits easy access for inspection and maintenance, although this is rarely necessary. PV arrays require very little, if any, maintenance over their long lifespan. The only maintenance I've had to perform on my PV systems is tightening conduit brackets and fixing one connector that pulled loose.

On the downside, ground-mounted arrays are not usually suitable for small lots in cities and suburbs. Trees and neighbors' homes often block the Sun in most backyards. Your own home, in fact, could block the sunlight striking a backyard-mounted array. Rural lots are typically better suited for ground-mounted arrays. That said, if a tracking array is installed, it needs to be positioned away from trees, barns, and other buildings to ensure dawn-to-dusk tracking.

FIGURE 8.20. Array Enclosed in Fence. This array at the First Methodist Church in Sedalia, Missouri, is protected by a chain link fence to keep kids and vandals out. Local building departments may require similar protection in urban and suburban settings. Notice the use of precast concrete blocks to anchor the array to the ground and ensure that it won't be blown away by the wind. Credit: Dan Chiras.

Ground-mounted arrays are also more accessible to vandals than building-mounted systems. In addition, precautions need to be taken to prevent livestock, such as horses and cattle, from using your $50,000 PV array as a scratching post. (This is usually solved by enclosing an array with an electric fence.) In more developed areas, local building departments may require fencing to keep kids out (Figure 8.20).

Ground-mounted arrays are also vulnerable to strong winds. An array becomes a large sail on windy days. Proper anchorage is critical.

Building-mounted PV Arrays

For many years, PV arrays were mounted on racks with back legs like the one shown in Figure 8.21. These racks allowed installers to set the correct tilt angle and are referred to as *elevated roof racks*. As solar became more popular, however, more and more building-mounted arrays were installed on standoff mounts like the one shown in Figure 8.22. Standoff mounts minimize the visual impact of PV arrays, making them more aesthetically appealing to homeowners, neighbors, and city officials. These are sometimes incorrectly referred to as *flush mounts*.

FIGURE 8.21. PV Array on Standard Rack. In early days of PV industry, modules were mounted on racks like the one shown here. They are designed to allow installers to mount modules at the proper tilt angle to maximize output, though they may not be the most attractive way to install a PV array.

FIGURE 8.22. Standoff Mount. Standoff mounts allow modules to be mounted parallel to the roof, permitting PV systems to blend in better, a feature that is desirable to homeowners, business owners, neighbors, and passersby. This array was installed by me, my business partner, and a former student on Victor Pipe and Steel in Winfield, Missouri. Credit: Dan Chiras.

FIGURE 8.23. Mounting Kit with Flashing. We used this roof mount with flashing on a job on a steel company in Winfield, MO. Flashings such as this help reduce the chances of leakage, giving the homeowner and the installer a little more peace of mind. Credit: Dan Chiras.

Traditional elevated roof racks required the same hardware that's used to install ground-mounted racks, discussed earlier: rails, L-feet, and fixed or adjustable back legs. Manufacturers have devised a number of attachments to secure elevated racks to roofs. They often include flashing that helps prevent leakage (Figure 8.23).

Standoff mounts require posts or standoff mounts to raise the racks off roofs, discussed shortly. Rails are attached to the standoffs by L-feet.

Installing a Building-mounted Elevated PV Rack

Installing both types of racks on roofs requires special skills and knowledge of roof design and construction. Workshop experience or the help of experienced friends can make the job a lot easier—and go a lot faster.

For best results, consider hiring a qualified installer—one who has installed a lot of systems. They can do the job in a fraction of the time and the result will probably be much better.

The first task when installing a system is to carefully measure the roof and locate the rafters to which you will secure the array. Rafters are usually set 24 inches (61 cm) on center. That means there's 24 inches (61 cm) from the center of one

FIGURE 8.24. Safety Harness. My business partner, Tom Bruns, roped in for safety on a two-story building in Missouri. Credit: Dan Chiras.

rafter to the next. In roofs with attics, it's a good idea to visually confirm this fact. While you are in the attic check out any abnormalities—rafters spaced differently, for example, or bowed rafters.

Once you have determined the rafter spacing, snap chalk lines on the roof to indicate the position of the front and back row of attachments—L-feet or other similar mounting hardware. Check your measurements two or three times *before* you start drilling holes in the roof.

Be sure to work safely. Consider using scaffolding to access the roof, rather than a ladder. Scaffolding is a lot safer. It will also make it easier to transport the modules onto the roof. Choose your shoes carefully. Sneakers with good tread often work well. Stiff work boots may not be as good on a roof. Stay off roofs when they're wet or storms are approaching. Metal roofs can be pretty slippery if covered with dust. On slippery metal roofs, we spray a sugary soft-drink like 7-Up. When it dries, it creates a slightly sticky surface that helps keep workers from sliding off the roof. Remember that roofs heat up on sunny days and hot asphalt shingles become soft and gooey and can easily be damaged by foot traffic. On hot days, my crew and I climb down by 10 or 11 AM, otherwise, we are sure to damage the shingles.

A full harness and a safety rope or two are vital (Figure 8.24). A lanyard connected to a worker's harness is used to connect workers to the safety rope strung across the roof. And, of course, never work when you are tired or impaired. Never work alone, either. Besides providing an extra set of hands, a helper provides another set of eyes and another mind that could help you avoid mistakes and ensure safe working conditions.

To secure an array to a roof, the L-feet are typically screwed into the rafters or top chords of ceiling trusses using lag screws. Lags should never just be screwed into the roof decking (Figure 8.26). It does not allow enough purchase to secure an array and prevent wind from lifting it off the roof.

Once you have snapped the lines, it is time to determine the position of the L-feet. L-feet are typically spaced 4 to 6 feet (1.2 to 1.8 meters) apart. How far apart depends on numerous factors including the height of the building, wind exposure (how far the rack is elevated by the back feet), and basic wind speed in your area. Spacing also depends on the steepness of your roof (roof pitch) and how deep the snow gets (snow load). It also depends on the ruggedness of the rail. Some are

Get It Right

Attaching an array via lag screws is a science. The length and diameter of the lag depends on the average wind speed in your area, which determines the uplift force. It also depends on the type of wood the rafters are made of—for example, southern pine vs. Douglas fir. That's because different species of trees create wood with different densities. Denser woods hold screws better and therefore resist winds better. Basically, the softer the wood, the deeper you must screw. The larger the diameter of the lag, the better it holds the array in place. Be sure to consult the rack manufacturer's guidelines when doing this. If you want to learn more, take a look at Jeff Tobe's piece in *Home Power*, Issue 161.

stronger than others. All these factors help designers place the L-feet or other mounting hardware so that they withstand uplift force created by the wind and the weight of snow on an array.

After the locations of all the L-feet have been marked using a grease pencil or chalk, it's time to start drilling. To confirm that you are on target, drill a tiny pilot hole through the roof with an ⅛-inch drill bit. This will allow you to determine whether you will hit a rafter or top cord. Once you are sure you are in the right spot, you may want to drill a slightly larger pilot hole, then drive a lag screw into the roof to secure the L-foot. Apply a dab of caulk to the hole and then caulk around the bottom of the L-foot.

If it's difficult to lag directly into framing members of a roof, for example, the rafters don't line up with the position of the attachments, wooden blocks can be installed between the rafters, provided you can access the attic. Screws are used to secure the blocks to the rafters and the L-feet are bolted (not screwed) into the blocks (Figure 8.26). Blocking is not possible in homes with vaulted ceilings, as there's no way to access the roof cavity.

When installing lag bolts or lag screws, be sure to apply a long-lasting caulk. I prefer polyurethane caulk. Caulk prevents rain or melting snow from penetrating the roof and soaking roof insulation. Get the best quality caulk possible. A few extra dollars spent on a high-quality caulk that ensures better protection against leaks could save you thousands in costly roof repair.

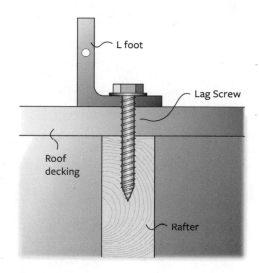

FIGURE 8.25. Lag Screw. L-feet or similar attachments are lagged into rafters or top cords of roof trusses with lag screws. Manufacturers specify how far apart the L-feet should be and the size and length of the lag screws, so be sure to check. Remember, too, you are only lagging into a 1.5-inch-wide framing member, so chances of missing it or running the lag at the edge are pretty high. Always drill a pilot hole first.

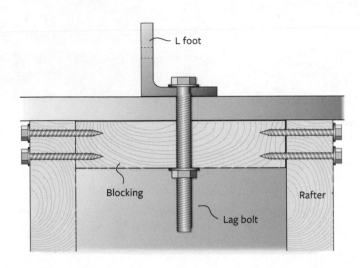

FIGURE 8.26. Lag Bolt and Block. Blocks securely attached to adjacent rafters can be used to attach an array to a roof. Credit: Anil Rao.

Modules are secured to the rack in one of two ways, either from the front (top-down) or from the back. Top-down mounting requires mounting clips like those shown in Figure 8.27. They are connected to slots in the rail via stainless steel bolts. In backside mounts, bolts run through the rail into holes in the frame of the module. This approach is rather difficult and slow, so, for ease of installation, most installers use top-down mounting clips, called *midclamps* and *end clamps*. Midclamps are also called *center clamps*.

When assembling a rack, be sure to use stainless steel nuts and bolts. They'll last much longer than their zinc-coated counterparts. Rusted nuts and bolts are unsightly and can be difficult to remove.

Once the modules are in place, they are wired in series using quick-connect adaptors on the module leads. These handy, waterproof snap-in connectors known as *MP-3 connectors* allow installers to connect one module to the next. The most positive and most negative wires from each string run to a combiner box at the array or in the inverter. That's where the series strings are wired in parallel. Parallel wiring, as you may recall, maintains the system voltage but increases the amperage.

FIGURE 8.27. (*a*) Midclamps Attached to Rail With WEEB. Midclamps, like the one shown here, connect the modules to the underlying rail. (*b*) A WEEB is used to bond the module frame to the rail. Credit: Dan Chiras.

When installing a PV array, be sure that the array and the racks—and the rest of the system—are properly grounded. Grounding is vital for safety and system performance. It protects users from shocks and helps protect equipment from damage. Shock may occur if insulation on a wire carrying current is damaged, that allows the internal conductor (aluminum or copper) to come in contact with metal, for example, the frame of a module or the metal case of a disconnect. If this occurs, the module frame or metal case will conduct electricity. If a person were to touch the metal, he or she could receive a shock.

Ground wires are typically 4 or 6 gauge bare copper wire. They connect all the metal components of a PV array, that is, the module frames and rails, to a ground-rod driven 8 feet (2.4 meters) into the ground (Figure 8.28). Copper is an excellent conductor of electricity. People are great conductors of electricity, too. We are filled with electrolyte in the form of watery fluids containing positively and negatively charged ions. Fortunately, electricity prefers the path of least resistance. And even more fortunately, copper ground wires provide a lower-resistance pathway for that

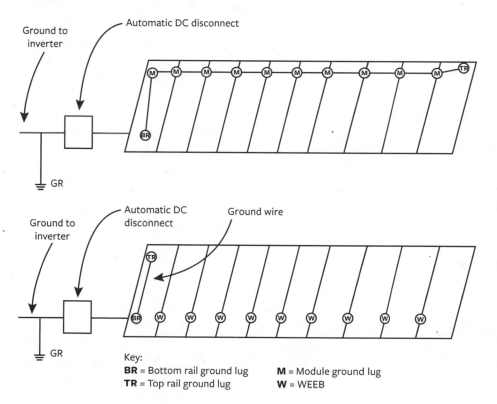

Key:
BR = Bottom rail ground lug   **M** = Module ground lug
TR = Top rail ground lug   **W** = WEEB

FIGURE 8.28. Grounded Array. Grounding a PV system is very important and most often not done correctly. (a) In this drawing, a ground wire is run from module to module but also connects to the top and bottom rails of each rack, forming a continuous ground. The ground wire terminates in a combiner box at a bus bar. A new ground wire runs from the combiner box to a ground-rod or to the inverter and main service panel, where the DC and AC systems can share a common system ground. Both are perfectly acceptable according to Code. (b) Shows how WEEBs can replace long ground wire runs. Credit: Forrest Chiras.

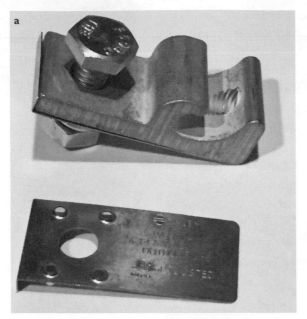

FIGURE 8.29. Ground Lugs. Ground lugs like these are (a) typically bolted to (b) rails and module frames with a WEEB to ensure a mechanical and electrical bond. Metal protrusions, or nipples, on the WEEB, as shown in (a), pierce the anodized aluminum coating of rails and module frames when a ground lug or midclamp is bolted tight. Credit: Dan Chiras.

FIGURE 8.30. WEEB. WEEBs like the one shown here help connect coated (anodized) aluminum of module frames to rails, ensuring that they are bonded and can freely conduct electricity, in case of a short circuit. Credit: Dan Chiras.

electrical current to travel to the ground than your body. If the system is not properly grounded, however, your body will provide an excellent path to ground.

When installing a PV array, you'll need to create a continuous ground. This can be accomplished in two ways. In days past, most installers ran ground wire from module frame to module frame, then from the top rail to the bottom rail to a combiner box (now incorporated in a rapid shutdown mechanism), and then to a ground-rod (Figure 8.28). To attach the ground wire to the rails and frames of modules, installers used *ground lugs* (Figure 8.29). The grounding wire was then run from lug to lug, then to the combiner box and then to the DC ground-rod located near the array (Figure 8.28).

To save time and materials, Wiley Electronics developed a product called a WEEB. It is a thin, flat piece of stainless steel that mechanically and electrically bonds each module to the rail at the mid clamps (Figures 8.29 and 8.30). These simple devices eliminate the need to install a continuous ground wire from one module to the next. They are part of a strategy called *integrated equipment grounding*. When WEEBs or similar devices are used, all that's needed to ground a PV system is a ground wire that runs from the bottom rail to the top rail, then to the DC ground-rod. Only one ground lug is required per rail.

WEEBs contain tiny protrusions, shown in Figure 8.30, that penetrate the protective coating (anodized) on aluminum rails and module frames, ensuring that electricity can flow from a module frame to the rail should a short occur.

Some companies manufacture midclamps that have serrated edges or teeth at key locations. They also permit integrated equipment grounding because the serrated edges pierce the anodized aluminum of the module frame and the rail, creating an electrical bond. For more on this topic, see Justine Sanchez's piece in *Home Power*, Issue 166.

Grounding an array can be tricky and it is often improperly executed. Be sure you consult with a knowledgeable professional. And, remember, all metal components of a PV system, including module frames, rails, and metal enclosures must be grounded. The ground system must connect all parts of the system, which means the ground wire in most systems must run from the array to the inverter to the main service panel. That said, things have gotten complicated with the transformerless inverters. When installing this type of inverter, there's no need to run a ground wire from the array to the inverter.

When installing an array, be sure to do a neat job. It looks much better! Loose wires should be attached to the rack with black (UV-resistant) zip ties.

Pros and Cons of Elevated Building-mounted Racks

One advantage of elevated roof-mounted racks is that they are, in most cases, relatively easy to install—at least easier than ground-mounted racks. To streamline installation and minimize time spent on the roof, some racks can be

Keep It Cool!

One of the secrets of generating the most electricity from PV arrays is to install them so that modules stay as cool as possible on sunny days. Even if it's a mild 77°F (25°C) outside, PV modules in direct sun will heat up—and substantially so. How they are mounted profoundly affects their output. The best options for keeping arrays cool are ground-mounted and elevated roof racks. Both of these arrangements permit better air circulation than standoff mounts. That said, the temperature of the modules will still be about 45°F to 54°F (20°C to 25°C) warmer than ambient (outside) temperature. Standoff-mounted PV arrays, which restrict air movement, will be considerably warmer—63°F to 72°F (35°C to 40°C)—than ambient temperature. The voltage of crystalline PV modules decreases about 0.3% per degree centigrade increase in temperature. Thus, the voltage of a module that's 72°F (40°C) warmer will be 12% lower than at standard test conditions (77°F [25°C]).

assembled on the ground and then carried onto the roof. The legs of the rack are then attached to the L-feet.

Another advantage of roof-mounted racks is that they allow installers to position PV arrays well above obstructions, such as trees or neighbors' homes. This helps boost an array's annual output.

Traditional elevated roof racks also permit air to circulate around an array, keeping it cooler and increasing its output. This type of rack also gives ready access the back of the modules, which makes wiring easier and makes it easy to troubleshoot modules, for example, to test the output of individual modules.

On the negative side, elevated racks are subject to the powerful forces of the wind. If not properly installed, a rooftop array can be ripped from its moorings by strong winds.

Roof-mounted racks also result in multiple roof penetrations. If not sealed properly, these can leak. Water leaking through the roof can dramatically reduce the R-value of ceiling insulation. (R-value is a measure of insulation's ability to resist heat flow.) Moisture also promotes mold growth and can discolor and damage ceilings. If the leak is persistent, it can cause framing members to rot.

All types of roof-mounted PV arrays, not just elevated racks, are also more difficult to access than ground-mounted or pole-mounted PV arrays. This makes it more challenging to adjust the tilt angle, brush snow off modules, or wash off dirt and bird droppings. That said, most roof-mounted racks are set at a fixed tilt angle and dust is rarely a problem.

Another downside of roof-mounting PV arrays—all roof-mounted arrays, not just elevated racks—is that they can limit an array's exposure to the Sun if the roof of a home doesn't face true south. This is not a shortcoming of the rack, but of the roof orientation. Some loss of output will result.

One of the most significant disadvantages of all roof-mounted racks is that they must be disassembled and removed when time comes to re-roof a house. This is costly and time-consuming work that should be carried out by a trained professional, not your local roofing contractor.

Asphalt shingles, found on the majority of homes in the United States live relatively short lives. This is especially true in areas that experience frequent, damaging hailstorms or areas exposed to intense sunlight or extreme temperature changes. So, when contemplating a roof-mounted rack, carefully evaluate the condition of your roof or hire a trustworthy roofer to give you an honest opinion. If your roof needs replacement, or will need replacement soon, do it before installing a PV system.

When building a new home or reroofing an existing home, I recommend installing a long-lasting roofing material. A metal roof should last for 50 years or more. Environmentally friendly roofing materials made from recycled plastic and rubber may also outlast standard asphalt shingle roofs by decades. At the very least, chose high-quality (long-lasting, architectural) asphalt shingles. They're more expensive, but could last 40 years.

One type of metal roof, called a standing seam metal roof, has an added advantage of "providing" a ready-made rack. True standing seam roofs are somewhat costly, but very long lived. Utilizing a clamp referred to as an S-5! clamp, the array can be mounted directly to the standing seams making for an installation that requires no roof penetrations.

Standoff Mount

Another building mount—the most popular of all options for homes—is the *standoff mount*. Shown in Figure 8.31, this rack allows installers to mount PV arrays parallel to the roof surface to reduce their profile and visibility. Because the array is mounted parallel to the roof surface, the modules are mounted at a fixed tilt angle corresponding to the angle of the roof.

In standoff mounts, the array is raised off the roof by about six inches (15 cm), sometimes more. (The more, the better from a performance standpoint.) The space between the roof and the array allows for some air movement, but nowhere near as much as other options.

"Standoff-mounted arrays are the most common, preferred, and least-expensive method for installing arrays as a retrofit or to existing rooftops," according to Jim Dunlop, author of *Photovoltaic Systems*, a superb advanced-level textbook on solar electricity. As noted earlier, the modules are attached to rails that are typically attached to metal feet (standoffs) lagged into the rafters (Figure 8.31). Grounding is best achieved with WEEBS.

FIGURE 8.31. Standoff-mounted PV Array. This home in a small town in western Missouri is equipped with (*a*) a roof-mounted PV array that blends in fairly nicely. (*b*) The modules are mounted six inches off the roof via a standoff mount. Credit: Dan Chiras.

FIGURE 8.32. Standoff-mounted PV Array S-5! Clamps. The S-5! clamp shown here allows an installer to attach a PV array to a standing seam metal roof. Rails can be attached to the clamps, then modules can be attached to the rails. Railless mounts are also possible with this mounting hardware. In such cases, the modules are mounted directly to the S-5! clamps, and no rails are used. Credit: S-5!

Installing a Standoff System

As just noted, standoff roof racks require standoffs lagged into the rafters or top chords of the roof trusses. They can also be lagged into blocks, as described earlier, if an installer can access the underside of the roof through the attic. Once the standoffs are attached, the rails are secured to them via L-feet. The PV modules are then placed on the rack and secured using midclamps and end clamps.

In buildings with standing seam metal roofs, rails can be attached to special clamps enthusiastically called S-5! clamps by the manufacturer. Rails can be attached to the S-5! clamps, and modules are then attached to the rails. To reduce material cost, some installers attach the modules directly to the S-5! clamps.

Shown in Figure 8.32, S-5! clamps connect to the seam in standing seam metal roofs, creating a secure attachment without penetrating the roof. Numerous types of S-5! clamps are now available.

Pros and Cons of Standoff Mounts

Standoff mounts are popular among homeowners and business owners primarily because they minimize the profile of an array, making it appear as if it were part of the building. This option is not only aesthetically appealing, it also reduces the wind's effect on a PV array. It's far less likely that a strong wind will dislodge a standoff-mounted array from its anchorage than it is for an array on an elevated rack, because the wind mostly passes over the top of the array.

While aesthetically appealing, standoff-mounted arrays have their downsides. One of them is that the pitch (slope) of the roof determines the tilt angle of the array. If the pitch of the roof does not correspond with the optimum tilt angle, the array's production could be seriously compromised. If the pitch is too shallow, for instance, wintertime production will be substantially lower than if the array were installed at the proper angle. For off-grid systems, reduced output, especially in the winter, can be devastating. To compensate, a homeowner needs to install a much larger and more costly PV array. If, on the other hand, the pitch is too steep, the array's output in the summer could be compromised.

Another disadvantage of standoff mounts is that they frequently result in higher array temperatures than rack-mounted or pole-mounted arrays. This reduces the array's annual output.

A third problem, encountered when installing any type of roof rack, is that the PV array must be removed when a roof needs to be reshingled.

Although standoff-mounted arrays are fairly easy to wire, it is more difficult to test the output of individual modules should you suspect a module or two are not performing well.

Standoff mounts are also not as well suited for use with microinverters as traditional roof-mounted racks. This is not to say that microinverters can't be installed. They can, but it's more difficult. In this case, you may want to consider installing a module with a built-in microinverter (called an *AC module*). Even so, there's another problem. That is, you can't see the LED idiot lights on the back of the microinverters that are used to troubleshoot problems. It's also a bit challenging to wire each microinverter to the special cable that the manufacturer provides.

FIGURE 8.33. Ballast Rack with Modules. These modules are mounted on a ballast rack. The racks are weighed down by cement blocks. This avoids penetration of roofs and is typically used on flat roofs of commercial establishments. Credit: Dan Chiras.

What's New: Ballast Racks and Railless Mounting

Homeowners or business owners with "flat roofs" (which are, in actuality, not flat, but only sloped enough to allow water to run off) usually prefer not to penetrate their roofs when having PV systems installed. In such cases, modules can be installed on weighted racks that sit on the roof. They are typically weighted down by cement blocks (Figure 8.33).

Ballast racks are made from lightweight aluminum or, in some cases, UV-resistant plastic. Most racks are equipped with ballast trays (or some similar structure) into which the cement blocks are placed. The trays raise the ballast off the roof to prevent damage. Modules are secured to the racks. Modules are bonded to the rack for grounding as shown in Figure 8.34.

Racks are typically installed in multiple rows, as shown in Figure 8.33. The individual rows of racks are connected to one another by a continuous ground wire that eventually runs to the DC ground-rod. To simplify installation, some manufacturers physically connect all of the rows of the ground-mounted rack (Figure 8.34). If the modules are electrically and mechanically bonded to the rack and the rack forms one continuous structure, grounding may simply require the installer to run a single ground wire from one point on the rack system to the ground-rod (Figure 8.35).

Ballast racks generally come preassembled, so installation is a snap. They do not require roof penetration. This is important on "flat roofs," which, though slightly sloped, don't always drain as well as more markedly sloped roofs. Lack of penetrations greatly reduces the chances of leakage.

There are some downsides to ballast mounts, however. One is that a great amount of ballast must be transported to the roof and carried to the ballast trays. Roofs must be strong enough to support the weight of the ballast (though this is usually not a problem).

Another major downside is that ballast racks are typically designed with a 10-degree tilt angle to prevent them from being lifted off roofs by wind. (Steeper angles would be possible, but you'd need more ballast.) The shallow angle won't affect the annual output of an array by much because it will enhance output in the summer. Winter production, however, will be much lower.

To simplify and reduce the cost of installation, rack manufacturers have developed railless rack systems. The first of these was the S-5! clamps for standing seam metal roofs, discussed earlier. Several manufacturers have created similar mounts for shingled roofs and metal roofs. These mounts are lagged into the rafters. They then attach directly to the frames of the modules, reducing cost and installation time. Figure 8.35 shows an example.

Railless mounting systems reduce the amount of aluminum required to install a PV array. Aluminum is a high-embodied energy material. Using a railless system, therefore, not only simplifies installation, and reduce costs, it results in less energy consumption and less air pollution.

FIGURE 8.34. Integrated PV Ballast Mount. Note how the racks from one row are connected to the rack in the next. Credit: Dan Chiras

Building-integrated PVs

In Chapter 3, I discussed the option for integrating PVs into buildings—for example, in roofs, walls, and glass. For most homeowners and businesses, there are three practical options: solar roof tiles, PV laminates on standing seam metal roofs, and solarscapes. Let's take a brief look at each one.

Laminate PVs for Standing Seam Metal Roofs

If your home or business is fitted with a standing seam metal roof, or if you're about to re-roof, you can install a product known as *PV laminate* (PVL) (Figure 8.36). Developed by United Solar Ovonics, which is no longer in business, PV laminate is fabricated by applying multiple layers of amorphous silicon to a durable, but flexible metal backing. It is adhered between the standing seams of standing seam metal roofs, and can be applied to new or existing roofs, though, it's best applied when building a new roof, for reasons discussed in Chapter 3.

Installing PV Laminate

PV laminate (PVL) comes in rolls. They're installed one at a time on sections of metal roofing. With the standing seam metal on sawhorses in a garage or shop, the ridge end of the roll is placed near the future ridgeline. The paper backing is then removed from the first 16 inches (40 cm) of the roll, exposing the super-strong adhesive that secures the PVL to the metal. The PVL is then secured to the metal, the remainder is then rolled in place and secured a little at a time as the PVL is unfurled.

When applying PVL, care should be taken to keep the laminate as straight as possible as it is unrolled. The adhesive is pretty strong. You can't lift and reposition PVL if you mess up. Be sure to avoid creasing.

Once the laminate has been laid out and secured, installers run a hard-rubber roller over it, pressing it in place. (It is the same device used to apply laminate to countertops.) The metal roof panel is then installed on the roof.

Once the roof panels are installed, the wire leads from the PV laminate are connected. Each section becomes a module. "Modules" are then wired in series. Leads are then run under the ridge cap if there's an attic or under eaves in buildings with vaulted ceilings. For ease of wiring and Code compliance, the manufacturer provides MC-4 connectors, a waterproof device that is difficult to take apart without a special tool.

FIGURE 8.34. Module Grounded to Ballast Rack. This rack system is integrally bonded. Notice how the rows are connected, forming a continuous metal structure that acts as a continuous electrical conductor that grounds all the rows together. A single ground wire is then run from the array to the ground-rod. Credit: Dan Chiras.

FIGURE 8.36. PV Laminate from Uni-Solar. PV laminate is applied directly to standing seam metal roof as shown here. Credit: National Renewable Energy Laboratory.

Pros and Cons of PVL

In skilled hands, PVL is fairly easy to install. It blends with the building and from a distance may not be visible at all. Unlike conventional crystalline PV, PVL is made from amorphous silicon, which is considerably less vulnerable to high temperatures—that is, their output doesn't decrease as much at high temperatures as monocrystalline and polycrystalline PVs. They're also not as sensitive to shading as standard PV modules. On the downside: because PVLs that are currently produced are less efficient than crystalline PVs, you may need twice as much roof space to produce the same amount of electricity. Standing seam metal roofing is also not as attractive, at least on homes, as other metal roofs. In fact, it is rarely used on homes.

Solar Tiles

Individuals interested in BIPV, may also want to consider solar roof tiles, like those manufactured by Sharp, Atlantis Energy Systems, and Kyocera. Solar tiles incorporate crystalline silicon cells in a tile that's attached directly to the roof. In one product, known as SUNSLATES, single-crystal PV cells are mounted on a fiber cement backing (Figure 8.37). Kyocera's and Sharp's solar tiles are made from polycrystalline PV cells.

Like solar shingles, solar tiles do not need to cover the entire roof. You can, for instance, dedicate a portion of a south-facing roof to them. The size of the system depends on the solar resource, roof pitch, shading, and electrical demand. As in all solar electric systems, the ideal location is a south-facing roof; however, solar tiles can also be mounted on roofs that face southeast or southwest. The decline in production will vary depending on how far off from true south the roof is oriented. Mounted on east- and west-facing roofs, the output could easily decline by 10% to 15% for shallower roofs. For really steep roofs, the decrease in performance could be as high as 50%.

Solar roof tiles overlap like shingles to shed water from the roof. Most products cannot be mounted on flat roofs. Roof pitch for SUNSLATES, for instance, must be at least 18 degrees (4/12 pitch). Be sure to check the pitch of your roof and manufacturers' recommendations before committing to solar roof tiles.

Solar roof tiles work well in a variety of climates, but perform best in sunnier locations. Be sure to check with the manufacturer for advice on the suitability of their product for your location. Also, be sure to check the compatibility of a solar tile with your roof. SUNSLATES, for instance, are "fully compatible with any other roof, be it tile, shake, metal or composite," according to the manufacturer. The manufacturer may also have recommendations for the best combinations. Atlantis Energy Systems, for instance, states that their product goes best with gray concrete roof tiles.

FIGURE 8.37. Sunslates from Atlantis Energy Systems. Sunslates from Atlantis Energy Systems are solar tiles that replace ordinary roof shingles. The top of this roof is fitted with solar tiles that generate electricity to help meet household demands. Credit: Atlantis Energy Systems.

Installing Solar Roof Tiles

Solar roof tiles are mounted on roofs over 30-pound roofing felt or similar roofing materials made from plastic. Each manufacturer provides directions. Kyocera's MyGen Meridian solar roof tiles are mounted on a metal rack that's flush mounted on the roof. SUNSLATES are mounted on a specially constructed wood framework—a series of vertical 2 × 2s secured to the rafters underneath (usually 2′ on center). Horizontal 1 × 4s are nailed to 2 × 2s about 12 inches apart. Hooks are then nailed to the 1 × 4s and the slates are hung on them. Creating a space beneath the tiles allows air to flow beneath the slates, cooling them and helping to boost output. Shaded roof areas are fitted with blank tiles.

Solar roof tiles are wired together to form series strings. SUNSLATES, for instance, are grouped in strings of 24. The home-run wires from each string are run to a combiner box or an inverter's combiner box. Installers run wires from a roof-mounted combiner box through a single roof penetration or may run the wire in conduit over the eave, then down the side of a building.

Pros and Cons of Solar Roof Tiles

Solar tiles offer many of the advantages of solar roofing products. The main benefit is aesthetics. They also have some of the same downsides, the most important of which is that they require a lot of connections. Many connections not only mean more work, they mean more resistance and lower efficiency. It also means that these arrays are more difficult to troubleshoot if something goes wrong. Designs I've seen also don't pack as many solar cells in as tightly as in crystalline PV modules. The net result of this is that you will need more roof space to produce the electricity required by your home or business.

Solarscapes

Another BIPV option designed to integrate solar electricity into buildings is *solarscaping*. In this approach, PV modules are incorporated into the roofs of structures such as pergolas, gazebos, and carports. PV modules can be used to create shade structures for decks, patios, or even hot tubs. The PV modules generate electricity while creating shade (Figure 8.38).

Solar shade structures are typically made from wood or steel and are designed and built so they blend in with the existing architecture and landscape.

Many installers use bifacial PV modules like those made by Sanyo and a few other manufacturers. Bifacial modules, discussed in Chapter 3, were developed in the mid-1970s for use in spacecraft. They contain PV cells encapsulated—both front and back—by glass. This allows the module to harvest solar energy from the front side and the backside. Sunlight reflecting either off the ground or off light-colored materials, such as metal roofing, concrete, snow, and water, strike the backside of the PV cells, generating additional electricity. The glass-on-glass construction of bifacial modules allows some light to filter through the array, creating "a soft light-and-shadow pattern on the surfaces beneath the array," according to Topher Donahue in an article on solarscaping in *Home Power* (Issue 122).

Pros and Cons of Solarscaping

Solarscaping is suited for new and existing homes and commercial structures. Like other types of BIPV, this approach helps minimize aesthetic concerns of conven-

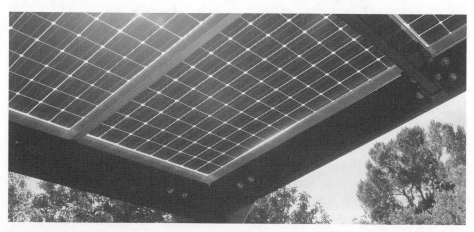

FIGURE 8.38. Solarscaping. PV modules can be used to create roofs of shade structures, a technique known as solarscaping. This photograph shows a portion of a carport made from PV modules by Lumos Solar. Credit: Lumos Solar.

tional PV mounting systems, specifically racks and pole-mounts. It therefore helps more people find a suitable way to incorporate PVs into their lives. The more options a company can provide their customers, the greater the chances they can meet their needs and the more solar electricity will be put in use.

The solarscaping idea was pioneered by a company called Lighthouse Solar. They manufacture structures for their customers in their workshop, then truck them to the site where they are assembled in one to two days, reducing time spent on a customer's property. You can contact them through their website at lighthouse solar.com.

Solarscaping allows placement of arrays in sunny locations on sun-challenged properties. In addition, it offers ease of access to the array and ease of installation compared to roof-mounted arrays. And, by avoiding installation on the roof, you won't have to worry about removing the array when the time comes to replace shingles. Like pole-mounted arrays, this approach results in a cooler and more energy-efficient array.

Conclusion

If you are interested in installing a solar electric system, it should be clear by now that you have many options. When considering options to mount an array, remember that one of your main goals is to produce as much electricity as possible. The accompanying sidebar, "Getting the Most from Your PV System" summarizes the many ways you can achieve this goal. Optimizing a PV array not only optimizes the value of your investment, it helps create a more sustainable supply of energy. But don't forget aesthetics and curb appeal. Although you may like the looks of a PV array mounted on a rack on a roof, neighbors might not be too keen. Future buyers may not like the look of it either.

Also, keep in mind the ease of installation. The more difficult and risky the installation, the more costly it will be. Although PV arrays require very little maintenance, access is especially important if you live in a dusty or snowy area. Arrays may need to be dusted off or have snow removed from time to time. Also, don't forget about winds and protecting an array from vandals and thieves.

Finally, solar electric arrays must be properly installed and must comply with all requirements of the local building codes. The electrical portion of the installation must comply with the National Electrical Code. (Article 690 of the NEC pertains to solar electric systems, but there are many other provisions under different sections of the Code that also apply.) If possible, you may want to consider making your array perform double-duty, generating electricity and providing shade, for example.

Getting the Most from Your PV System

To produce as much electricity from your PV system:

1. Locate your PV array in the sunniest (shade-free) location on your property, orienting the array to true south, setting modules at the optimum tilt angle or installing on an adjustable rack or tracker. Ensure access to the Sun from 8 AM to 4 or 5 PM!

2. Install a Maximum Power Point Tracking controller (for battery-based systems) or inverter for grid-connected systems to ensure maximum array output.

3. Keep modules as cool as possible by mounting them on a ground-mounted rack or pole or an elevated roof rack. In hot climates, install high-temperature modules, that is, modules that perform better under higher temperatures.

4. Install high-efficiency modules if roof space or rack space is limited.

5. When installing multiple rows, be sure to provide a sufficient amount of distance between the rows to avoid inter-row shading.

6. Select modules with the lowest rated power tolerance. Power tolerance is expressed as a percentage, which indicates the percent by which a PV module will overperform (produce more power) or underperform (produce less power) the nameplate rating. Look for modules with a small negative or a positive-only power tolerance.

7. Install an efficient inverter. Use the weighted efficiency as your guide rather than maximum efficiency. Weighted efficiency is a measure of the efficiency under typical operating conditions.

8. Keep your inverter cool. If installed outside, be sure the inverter is shaded at all times. If installed inside, select a cool location. Be sure air can circulate around the inverter.

9. Decrease line losses by installing a high-voltage array. Reduce the length of wire runs whenever possible. Use larger conductors than required by Code to reduce resistance losses. Be sure all connections are tight to reduce resistance and conduction losses.

10. Keep your modules clean and snow free. If you live in a dusty environment with very little rain, periodically wash your modules. Remove snow from modules in the winter, but brush it off gently. Don't use any sharp tools to remove snow. Mount the array so it is easily accessible for cleaning and/or snow removal.

11. For battery-based systems, maintain batteries properly. Periodically equalize batteries and fill flooded lead-acid batteries with distilled water. Recharge promptly after deep discharges and install batteries in a warm location so they stay at 75°F to 80°F, if possible.

CHAPTER 9

Final Considerations: Permits, Covenants, Utility Interconnection, and Buying a System

To those who are enamored by the idea of generating their own electricity from the Sun, there are few things in the world more exciting than turning a PV system on for the first time and watching the meter run backward—meaning you are not only producing all the electricity your home needs, you are producing a surplus that you can use later. I've installed over two dozen PV systems for myself and clients and still get a thrill when we "fire up" a PV system. It's an even greater thrill when you receive your first utility bill and discover the only fee you have to pay is the monthly service charge.

Although you may encounter a few obstacles along the way, in most instances, the path from conception to the completed installation is fairly straightforward—and it is getting easier as more and more utilities are becoming familiar with solar electric systems.

The sidebar on the following page summarizes the steps you or your installer must take, in the order in which they must be completed. As you can see, we've already discussed steps 1 and 2 and details of others. In this chapter, we'll explore the remaining steps. I will begin by discussing permits, then look at restrictive covenants imposed by some homeowners and neighborhood associations. I will then discuss interconnection agreements required for grid-connected PV systems and what kind of insurance. I'll end with some advice on buying and installing a PV system.

Steps to Implement a PV Energy Project

1. Determine your home or business' electrical consumption and consider making efficiency improvements to reduce the PV system size and cost.
2. Assess the solar resource.
3. Size and design the system.
4. Check homeowner association regulations; file necessary paperwork for permission to install the system, if required.
5. Apply for special incentives that may be available from the utility or your state or local government.
6. Check on building permit requirements; file a permit application.
7. Check on insurance coverage.
8. Contact the local utility and negotiate utility interconnection agreement (for grid-connected systems).
9. Obtain permit.
10. Order modules, rack and balance of system.
11. Install system.
12. Commission—require installer to verify performance of the system.
13. Sign interconnection agreement.

Permitting a PV System

After you or your installer have designed a PV system to meet your needs, you'll need to contact your local building department—managed by the city, town, or county—to determine if they require a permit. In most cases, an electrical permit will be required. If, however, you live in a less-regulated state like Missouri, you may find that only a handful of counties require permits, often only those that are home to major cities. To find out where your building department is located, ask a builder or an electrician or call local government. Once you locate the permitting agency, ask for a permit application. If you are installing your own system, they will often guide you through the process. If you hire a professional, he or she will usually secure a permit.

Building departments are known as AHJs, *authorities having jurisdiction*, and they control all aspects of construction in cities, towns, or counties. AHJs are granted the authority to administer, interpret, and enforce building codes by local governments.

Building codes are a detailed set of regulations that apply to how buildings are constructed, modified, and repaired. Building codes set the rules by which the various trades must operate. They stipulate equipment and materials that can be used and how they must be installed. For example, they stipulate the way PV systems must be wired and the use of safety measures such as overcurrent protection (circuit breakers) and disconnects. Local trade licensing codes (separate from building codes) may also stipulate who can perform certain tasks, for example, licensed electricians or plumbers. Building departments may also stipulate setbacks—that is, how close a PV system may be installed to a public right of way or a neighbor's property line.

Building departments may require an electrical permit for a PV installation and a letter from a structural engineer certifying that your roof is strong enough to support the PV array. They may also require a structural permit for ground-mounted racks. Building departments also conduct inspections at certain stages during installation. For a roof-mounted PV system, only one inspection, the final inspection is typically required. For ground-mounted PV arrays, the building department may want to inspect the trench in which you run your wire from the array and the holes you dig for the concrete piers and pads that support the steel rack.

When a project is completed and passes all required inspections, the local AHJ grants a certificate of occupancy, in the case of a building, or a certificate of approval, in the case of a PV project. Bear in mind that building permits will be required for off-grid systems if you live in a city, town, or county that has adopted the National Electric Code.

Although AHJs can establish their own regulations for construction and remodeling, the process is expensive, time consuming, and extremely costly. (In the United States, only Chicago and New York City have developed their own building codes.) Because of the time, labor, and expense required to create a comprehensive building code, most AHJs have adopted model building codes, created by independent organizations. In the United States, most jurisdictions have adopted the *National Electrical Code* (NEC) to govern all electrical work on homes and businesses, including renewable energy systems.

The National Electrical Code was developed by the National Fire Protection Association's Committee on the National Electrical Code. This group consists of 19 code-making panels and a technical correlating committee. Article 690 of the NEC applies specifically to the installation of solar photovoltaic systems. As noted in the last chapter, other articles of the NEC also apply to solar installations—so you've really got to know what you are doing.

Although states, cities, and counties typically adopt a model code, they are given the authority to modify the code to meet their needs and the needs of local builders and electricians. For example, AHJs may waive certain requirements or permit alternative measures or equipment that ensures the same safety standards. When it comes to solar installations, AHJs typically rely entirely on the NEC.

The National Electric Code also lists the equipment that can be used in PV installations. If a component is not listed, it can't legally be installed in areas where building codes are enforced.

Equipment is listed for PV installation by independent testing laboratories, the largest of which is Underwriter's Laboratories (UL). It tests equipment like modules and inverters, and, if the equipment passes a set of criteria, it lists (certifies) the product as meeting its standard. The UL listing is shown on the nameplate of the product. UL-listed products may comply with US or Canadian safety standards, or both.

While the UL subjects PV equipment—including modules, inverters, and charge controllers—to rigorous testing, they don't test every piece of equipment that rolls off the manufacturer's assembly line. Rather, they test representative samples and then routinely inspect production facilities for ongoing compliance with UL standards.

When reviewing permit applications, plan examiners of the local AHJ check to see that all equipment is UL-listed. That's why it is a good idea to send spec sheets for all the major components of a PV to the AHJ when you submit your permit. Building department inspectors will verify that the equipment that was specified in the plans submitted with the permit applications was actually installed—if they perform a very detailed inspection of a PV system, which most don't.

Inspectors are licensed and certified by either the state or local government and are supposed to have a thorough knowledge of the Code. In rural areas where funds are limited, one inspector may be licensed to inspect electrical, plumbing, and structure. You may have to spend some time educating them on PV. If you installed your own system, it is a good idea to be on hand when they inspect a system to address concerns.

Although there's a lot of grumbling about building codes and permits among builders and installers, codes serve an important purpose. They ensure that all building projects, including the installation of PV systems, are safe for the immediate occupants of a home or business—as well as for future residents of a home or employees of a business. The NEC, for example, protects against potential shock and fires caused by electricity. If you install a PV system that is not in accordance with the Code, and it causes a fire that destroys your house, your insurance company may deny your claim. The NEC provisions that apply to PV systems also help ensure the safety of electricians who may come to work on your system and utility company employees (notably, line workers) in the case of utility-connected systems. In short, building codes are society's way of ensuring the safety and well being of present and future generations.

Although the requirements of local building codes can be daunting to the uninitiated, professional installers should be intimately familiar with them.

Securing a Permit

To start the process, as noted above, you or your installer must submit a permit application. Permits for PV systems are pretty simple. They encompass electrical wiring and all the electronic components you'll be installing, such as inverters, charge controllers, and safety disconnects. Your application may also be required to describe how the array is mounted, so the permitting authority can be sure that your roof will support a roof-mounted array and that it will be securely attached to the roof so it won't blow away in the wind. (More on this shortly.)

To determine if a permit is required, even for off-grid systems, give your local building department a call. If a permit is required, they'll outline the procedure, indicate the cost, and provide the form. They will also indicate all the supporting material you'll need to provide as part of your application.

Applications may require a site map, drawn to scale, that indicates property lines, streets, and the proposed location of the array (Figure 9.1). If required, this drawing should also indicate the location of other components of the system, for example, the inverter, combiner box, disconnects, and electric meter (in grid-connected systems). A professional installer will prepare a site map for you. It's part of his or her service.

Although building permit applications may not require a site map, they all require a one-line drawing. As shown in Figure 9.2, a single-line drawing shows all the components of the system, including the PV array. It also lists the specifications of each component and shows wire runs, indicating the size of wire required to comply with the NEC. In some jurisdictions, AHJs may require more complex three-line electrical diagram. These drawings shown all the wires, including ground wires.

One-line drawings not only indicate all of the components of PV systems, they show how the system will be connected to the utility grid—for example, whether it will be connected via the main service panel.

You may also need to provide a description and/or drawings of the array-mounting design and the materials when applying for a permit. If you're mounting an array on a pole-mount, you may need to provide drawings of the foundation and

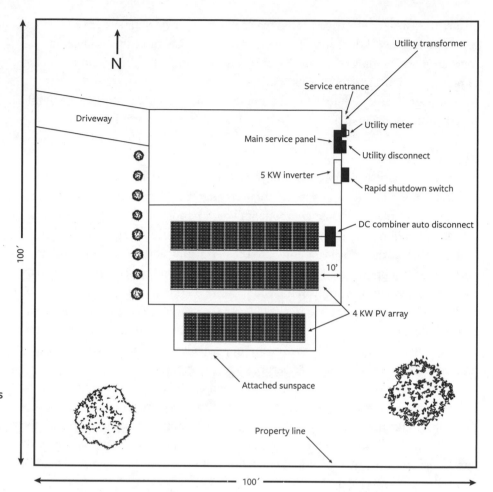

N

Utility transformer

Service entrance

Driveway

Utility meter

Main service panel

Utility disconnect

5 KW inverter

Rapid shutdown switch

DC combiner auto disconnect

10'

4 KW PV array

Attached sunspace

Property line

100'

100'

FIGURE 9.1. Sample Site Map. A site map like this one shows the location of the array and important equipment. It may be required by your building department. Credit: Forrest Chiras.

pole-mount. If you are mounting an array on a roof, some AHJs ask you to include information on the age of the roof, type of shingles, the pitch, and the size and spacing of the rafters. You may also need to provide information on the weight of the array and the method of securing it to the roof. You may need to provide details on waterproofing the attachment as well, although that's unlikely.

Because building departments may not want to make a determination as to the strength of your roof, they will most likely ask you to hire a professional engineer to review and stamp your plans. An engineer's stamp ensures that the roof can support the array and that it will remain intact under local wind and snow conditions.

Even if your local building department does not require an electrical permit, which is rare, the local utility will ask for a one-line drawing and spec sheets for

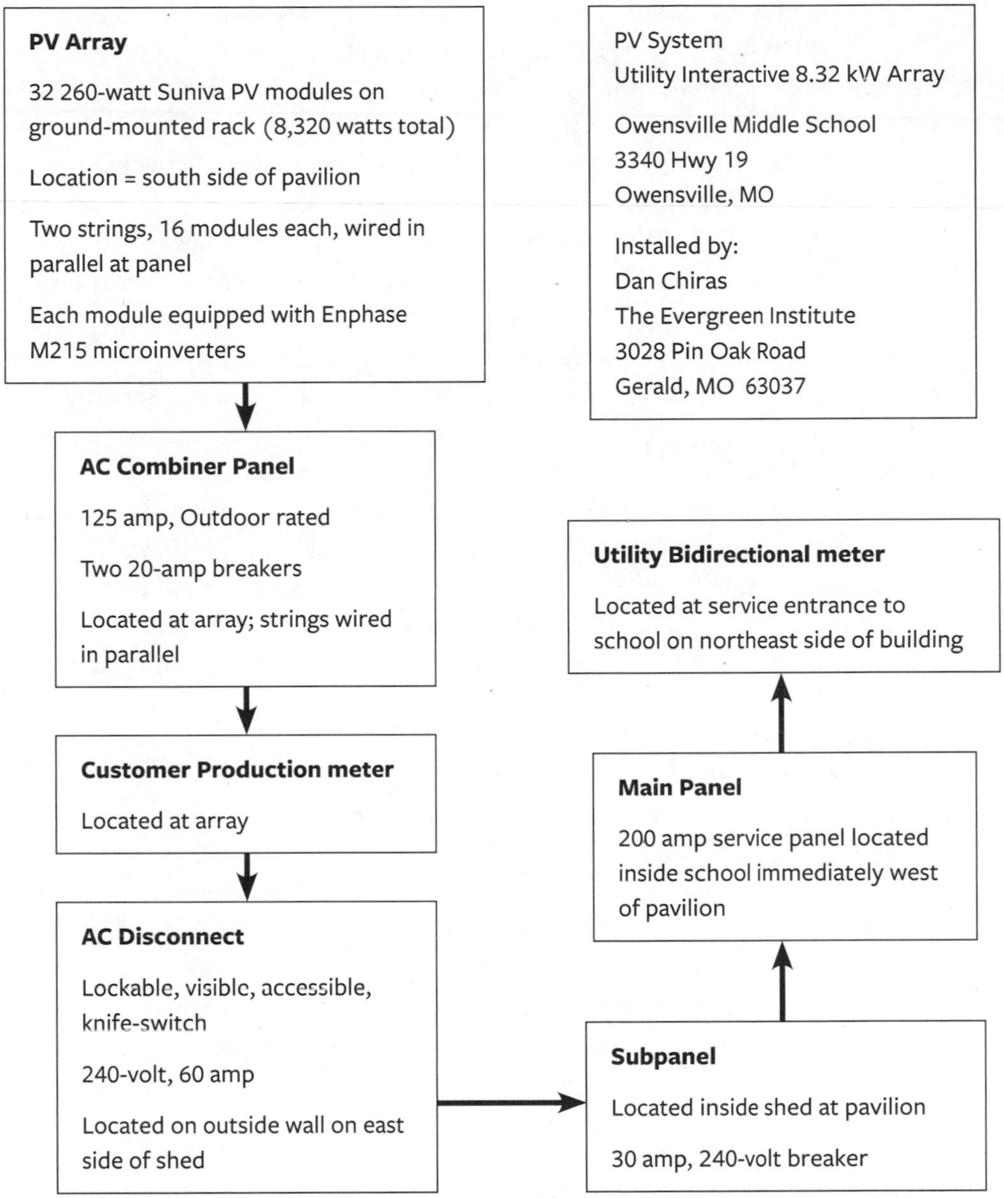

PV Array

32 260-watt Suniva PV modules on ground-mounted rack (8,320 watts total)

Location = south side of pavilion

Two strings, 16 modules each, wired in parallel at panel

Each module equipped with Enphase M215 microinverters

PV System
Utility Interactive 8.32 kW Array

Owensville Middle School
3340 Hwy 19
Owensville, MO

Installed by:
Dan Chiras
The Evergreen Institute
3028 Pin Oak Road
Gerald, MO 63037

AC Combiner Panel

125 amp, Outdoor rated

Two 20-amp breakers

Located at array; strings wired in parallel

Customer Production meter

Located at array

AC Disconnect

Lockable, visible, accessible, knife-switch

240-volt, 60 amp

Located on outside wall on east side of shed

Utility Bidirectional meter

Located at service entrance to school on northeast side of building

---

Main Panel

200 amp service panel located inside school immediately west of pavilion

---

Subpanel

Located inside shed at pavilion

30 amp, 240-volt breaker

FIGURE 9.2. One-line Drawing of Electrical System. This one-line drawing shows the layout of a PV system, including all the equipment that will be installed and pertinent details about that equipment. Wire sizes are also included. Building departments always require these drawings, and utilities may request one as well. Be sure to check your AHJ and utility first to see the level of detail they ask for. In my experience, the more detail, the better. Credit: Dan Chiras.

modules and inverters. Because rural utilities often don't have the personnel trained to assess PV systems, utilities typically ask the installer to submit a pre-construction certification—a statement from a licensed electrician or engineer who states that the system was designed in compliance with the NEC.

If you are hiring a professional installer, he or she should file the permit and provide engineering reports, if required. Be sure they do this. If you're installing a system yourself, this task rests on your shoulders. If that sounds like too much, you may want to consider buying a kit, which includes all of the components you'll need, including diagrams of the system and specifications of the components. Kits make it easier on the do-it-yourselfer. Be sure all the components, like inverters, are produced in the United States or Canada, or are at least easily serviced. You don't want a system with an inverter or other equipment that comes from a foreign manufacturer that doesn't offer you any tech support or service.

After your application is submitted, a plan examiner in the building department will review the application and accompanying materials, usually within a few days or weeks, although the process can take several months, depending on the workload and staffing of the AHJ. If everything is in order, you'll receive a permit, an official approval for you to commence construction. (Never order equipment or start work until you've received your permit!)

If your permit application doesn't meet the Code, the plan inspector will mark up the schematics—it's called *redlining*—or provide written notice indicating the problem or problems and required changes. Once the problems have been resolved, you must resubmit the permit application.

Fees for permits for residential PV systems usually run from $50 to $1,500, depending on the jurisdiction. Some jurisdictions charge a flat fee; others charge a fee based on the size of the system. Permits must be posted on the site, usually in a window. Expect an electrical inspector to visit your site at the end to check wiring and warning labels. (Warning labels are required on disconnects and other components of PV systems). The inspector may check on wire sizes, connections, overcurrent protection, and grounding.

All equipment must be readily accessible to inspectors and must be installed according to Code with proper clearances (to ensure room to work on the equipment). This is known as *working space*. You'll also need to be sure that no other wiring, plumbing, or ductwork is within a certain distance of installed equipment. You or the installer must call the AHJ to schedule all inspections, usually 24 to 48 hours in advance.

If you fail an inspection, you'll need to fix the problem and arrange a follow-up inspection. You may have to pay for follow-up inspections, too, although the cost of follow-up inspections is usually included in a one-time permit fee.

When applying for permits yourself, be sure you know what you are doing or hire a professional installer who does! Be courteous and respectful in all your dealings with the building department. If you are feeling irritated at having to file and pay a fee, leave your angst at home. The same goes for dealing with inspectors. Inspectors have a tough job. They deal with difficult electricians and builders day in and day out and often show up to a site in a sour mood. Be polite and respectful, even friendly—even if they fail to pass your installation. Ask for explanations on ways to correct mistakes. They're usually happy to help out.

When applying for permits or dealing with inspectors, however, expect to be treated by building department officials in a reasonable and timely manner. A pugnacious attitude and list of demands are not well received. "From the official's perspective, there is no bigger turn-off than having to deal with an arrogant know-it-all," notes Mick Sagrillo, a veteran wind system installer. "On the other hand, don't be intimidated by resistance when dealing with building code officials."

While we are on the subject, avoid the tendency to bypass the law—that is, to install a PV system without a permit. Do not, under any circumstances, try to sidestep this requirement. The consequences are too great. Municipal governments have the authority to force homeowners to remove unpermitted structures, even entire homes. Be sure to sign an interconnection agreement with the local utility, too. With all the sophisticated metering and monitoring in place, the utility will know if you are backfeeding electricity onto the grid. If you live in a county that does not require permits, be sure to install your system to Code. Lack of enforcement is no excuse for shoddy, unsafe installations.

Permitting a PV system may take several months, so submit your application well in advance of the date you'd like to install the system, *and don't buy a PV system until your permit has been granted.* And, as a final note on the subject, keep a copy of the certificate of approval. You may need it in the future if you file an insurance claim.

Covenants and Neighborhood Concerns

Restrictive covenants can create a huge obstacle to a PV system installation. Even with a permit from the local building department, restrictive covenants can block an installation.

Covenants are agreements incorporated into the document under which sub-divisions are created. Covenants and neighborhood association rules are found in many neighborhoods and subdivisions throughout North America.

Covenants require or prohibit certain activities. These covenants "run with the land," meaning they apply to the original owner and subsequent owners of the property as well. Some covenants give the neighborhood association the right to create and enforce additional rules and architectural standards.

Restrictive covenants and neighborhood association rules may address many aspects of our homes—from the color of paint we can use to the installation of a privacy fence. Some neighborhood associations even ban clotheslines. Some expressly prohibit renewable energy systems, such as solar hot water systems and solar electric systems.

Restrictive covenants are legally enforceable, and the courts have consistently upheld their legality. To see if you will be prohibited from installing a PV system, you need to review both the restrictive covenants and neighborhood association rules, if they exist. Contact your neighborhood association to see what rules apply.

If your subdivision has restrictive covenants or architectural standards, you need to apply for permission to install a PV system. The application is often a written letter with a drawing or two. Precedence can help. In other words, if someone else has installed a similar system, even without permission, it's easier to obtain approval.

Even in the absence of restrictive covenants, you might want to consider your neighbors. Many people feel very strongly about protecting the aesthetics of their neighborhood—and the resale value of their home. Be sure to discuss your plans to install a PV system with your closest neighbors, long before you lay your money down, especially if you live in a covenanted community.

Don't expect your neighbors to be as enthusiastic about a PV system as you are. Not everyone is enamored of renewable energy. Some people have a knee-jerk reaction against it. Why?

Many people fear the unknown. Others object to anything new or "environmental." Others may have distorted notions of what a solar system might look like. They may have seen an ugly system at one time and assume that every system is going to look the same. Still others may object out of spite or because of unresolved anger.

To help allay fears and win over your neighbors, you may want to show your neighbors pictures of the type of PV system you're going to install. If you are good at Photoshop, take a photo of your home from their house and place a picture of the PV system on your roof or on your property so they can see what it will look like. If you are thinking about installing a grid-connected system, let them know that you'll be supplying part of their energy, too. Giving neighbors advance notice, answering questions, and being responsive to their concerns is the best way to ensure peace in your neighborhood.

If you're building a new home in a historic district or adding a PV system to an existing home in a historic district, you may have a battle on your hands. A building-integrated PV system may be the only possible option, and even that may not meet approval. However, if you can locate the array where it is not visible from the street, your chances for approval should increase.

Covenants can be a pain. When I published the first edition of this book, two renewable energy-friendly states—Colorado and Wisconsin—prohibited homeowners associations from preventing homeowners from installing solar and other renewable energy systems. Today, about half the states in the nation have similar laws. If you're in one of those states, count yourself lucky. If not, you might want to work to pass such legislation in your state.

Connecting to the Grid: Working with Your Local Utility

When installing a grid-connected system, you'll need to contact your local utility company early on—*well before* you purchase your equipment—to let them know what you're planning and to secure a copy of their interconnection agreement. It's a good idea to call your utility and speak to the person in charge of grid interconnection. Be courteous. These people can be great allies! My local utility—Three Rivers Electric—is a dream to work with. They are cooperative, friendly, and responsive to my concerns. We've become good friends over the years.

One of the main concerns all utilities I've dealt with have is that a PV system won't backfeed electricity onto the grid if it goes down. Because UL 1741-listed grid-tied inverters shut down automatically when the grid goes down, all the utility generally needs is documentation showing that your inverter is UL 1741 listed. (Inverter manufacturers put this listing on their spec sheets; some manufacturers issue certificates of compliance with UL 1741 that you can download from their websites.)

Utilities in all but California and Colorado will also ask you about the location of the manual AC disconnect—the utility disconnect. As noted in Chapter 5, most utilities require the installation of visible, lockable AC disconnects—that is, manually operated switches that allow utility workers to disconnect your system from outside your home in case the grid goes down and they need to work on the electrical lines. As you know by now, because grid-connected inverters automatically disconnect from the grid when they sense a drop in line voltage or a change in frequency, this requirement is unnecessary. Regardless, it may still be required by your local utility. There's no way around this requirement.

Utility interconnection agreements also require information on the size of the system and the inverter make and model. In addition, the interconnection agreements will spell out their net metering policy, which is dictated, for the most part, by state law. This agreement will explain how the utility handles net excess generation—that is, whether they have annual or monthly reconciliation and how much they reimburse customers for surplus electricity.

Interconnection agreements may also suggest the purchase of a liability insurance policy. You will probably be shocked when you see that many companies suggest that homeowners secure $1,000,000 liability insurance policies. It's meant to protect the utility from damage your PV system might cause to their workers, their system, or your neighbors—all of which are deemed highly unlikely by professionals. Few people secure policies of this size, as you shall soon see.

Interconnection agreements also include a pre-construction document that stipulates that the system will be installed according to Code. They also include a post-construction agreement, which is a form submitted to them after the system has been installed, saying that the PV system was installed according to Code. Both these documents need to be signed by an electrician or engineer, though finding one who understands Code requirements for a PV system can be difficult in rural areas and small towns.

While all this may sound daunting, remember: if you are hiring a professional installer, he or she should take care of this paperwork for you. If you are going it alone, you'll be happy to know that the forms are relatively easy to fill out. You will, however, have to hire an electrician or engineer to sign the pre- and post-construction documents.

Some utilities charge an interconnection fee. This fee may be based on the size of the system, or it may be fixed—with one price covering all systems. Fees typically range from $20 to $800. They cover costs borne by the utility, such as the cost of inspecting a PV system after it is installed, and, more likely, the cost of installing a bidirectional meter. Be sure to ask about interconnection fees upfront and be sure the contract is clear about any additional fees that may be charged to you over the course of the contract.

As when working with a building department, keep in mind that being overly assertive or an arrogant know-it-all won't help matters. Your utility can make the approval process very fast or unbelievably difficult. "It has all the marbles, and it has the bag, too. And it knows it," Sagrillo notes. If you run into problems with the utility and feel you are in the right, your best bet is to work through your state's public utility commission.

If you go into the process armed with the proper knowledge and equipped with a good attitude, you should expect smooth sailing. Invite the utility to bring staff out to inspect your PV system. Many could become believers once they witness your PV system generating electricity and feeding it onto the grid. I've hosted dozens of workers from utilities over the years—as well as building department officials—who wanted to learn about PV systems.

Making the Connection

Once a grid-connected PV system has passed final inspections by the local AHJ, you or the AHJ must contact the utility. At that point, they will briefly inspect the system, change out the meter, if necessary, and sign the interconnection agreement, granting you the right to connect to the grid. Only then can you flip the switch and start generating electricity. In some cases, utilities may test a PV system to be sure the automatic disconnect functions as it is supposed to.

Whatever you do, don't switch a PV system on *before* the utility has approved your system for interconnection. While it may take a week to ten days to get the utility to show up at your home, and it is tempting to start producing electricity before they get there, please resist the nearly overwhelming temptation. That said, an installer can connect for brief periods to test a system to be sure everything is working. Even then, I like to phone the utility first to request permission. I've never been denied. Utilities realize that an installer cannot determine whether a system is working correctly without momentarily going on grid. However, once the installer shuts the system off and leaves the premises, resist the temptation to flip the switch to fire that system up. The utility knocked on the door of one of my eager customers the next day—and they were hopping mad!

Insurance Requirements

Besides permits, inspections, and utility approval for interconnection, if the system is grid-connected, you will need to secure insurance for your system. Two types of insurance are required: property damage which protects against damage *to* the PV system, and liability insurance to protect you against damage *caused* by the system. Both are provided by homeowner's or property owner's insurance policies.

Insuring Against Property Damage

For homeowners, the most cost-effective way to insure a new PV system against damage is under an existing homeowner's insurance policy. Businesses can cover a PV system under their property insurance, too.

When installing a PV system on a home or outbuilding—or even on a pole-mount—contact your insurance company to determine if your current coverage is sufficient. If not, you'll need to boost the coverage enough to cover replacement, including materials and labor. If you are installing a PV array on a pole, be sure to let them know that the PV system should be insured as an "appurtenant structure" on your current homeowner's policy. This is a term used by the insurance industry to refer to any uninhabited structure on your property not physically attached to your home. Examples include unattached garages, barns, sheds, satellite dishes, and wind turbine towers.

Insurance premiums on a homeowner's insurance policy fall into two different categories, each with differing rates. Your home is assessed at a higher value than an unattached garage or a storage shed or your tower. This is because people's homes are more lavish than most garages and sheds, and contain myriad personal possessions, furniture, and clothing not typically found in other structures.

Appurtenant structures, on the other hand, are assessed and charged at a lower rate. Insurance companies usually base premiums for appurtenant structures on the total cost of materials plus the labor to build the structure. It is best, although a bit more expensive, to insure a PV system at its full replacement cost—not a depreciated value. PV systems can easily last three decades—or more. Any PV system should have insurance coverage that includes damage to the system itself from "acts of nature," such as lightning or wind. While most PV systems are designed to withstand 100-plus mile-per-hour winds, tornadoes or hurricanes can destroy them, just as they would any other structure in their path. You may want to consider riders that cover fire, theft, vandalism, or flooding.

Another "act of nature" of concern is direct or nearby lightning strikes. When lightning strikes nearby, it sends out a magnetic pulse. The pulse passes over wires in the ground and the metal parts of the array where it can generate a sizeable electrical current. Grounding systems help dissipate this charge, but even so, a high amperage surge can burn out equipment.

Not previously discussed, it is important to install lightning arrestors/surge protectors in PV systems. For example, a lightning arrestor/surge protector should be installed in the DC combiner box or DC disconnect of conventional systems. After losing an inverter to a nearby lightning strike, I always install a lightning arrestor in my string inverters as well. A lightning arrestor and voltage surge arrestor should also be installed on the utility side of the inverter to provide additional protection. They protect against high-amp surges from the utility, which can be caused by lightning strikes or malfunctioning equipment. In systems that incorporate micro-inverters, we like to install the lightning arrestor/surge protector in the AC combiner panel.

While conventional lightning arrestors will not guarantee that your system is safe from lightning, it may reduce the damage from a rare direct or (more likely) nearby lightning strike. Fortunately, there are some new-generation lightning arrestors on the market from MidNite Solar that are much more likely to succeed in protecting a system than conventional lightning arrestors. Either way, in the eyes of the insurance company, if you have installed one, you have taken prudent measures to protect your system. If your inverter is fried by lightning, you should be able to collect from the insurance company.

Improperly wired PV systems can cause house fires. If you have a fire in the house caused by some funky wiring in the system, your claim may be denied and your policy canceled.

Theft and vandalism of a PV system is unlikely—unless the modules are mounted on the ground in high-crime areas or neighborhoods with mischievous children. Flood insurance is a nationally administered program to protect primary dwellings. Costs can be exceptionally high for a home located along a coastline or in a floodplain—that is, alongside a stream or river with a penchant for flooding. If you live in a floodplain, obtain an insurance estimate before beginning construction.

Insuring a PV system is relatively inexpensive. While home insurance coverage should cover appurtenant structures, added insurance may be required. It can be purchased for an additional premium.

Liability Coverage

Liability insurance protects against possible damage to others caused by a PV system. Although I have never heard of any such damage occurring, liability insurance is worth having.

Liability insurance covers possible claims from damage to a neighbor's electronic devices from a grid-connected PV system. (For example, if your system sends power onto the line that somehow, magically, damages electronic equipment in a neighbor's home.) It also covers personal injury or death of employees due to electrical shock from a system when working on a utility line during a power outage. Even though the likelihood of this is nil, because of the automatic disconnect feature built into inverters, utilities may recommend this coverage. However, as noted earlier, they typically suggest liability coverage way in excess of what's reasonable.

Liability coverage is part of a homeowner's or business owner's policies. In most places, liability coverage for homes provides protection from $100,000 to $300,000.

In addition to liability insurance, utilities may require you to indemnify them from potential damage from your PV system. For more advice on insurance, and specifically on working with insurance agents, see the accompanying sidebar, "Insuring Your PV System."

Insuring Your PV System

Insuring a PV system should be pretty straightforward; however, not all insurance agents understand renewable energy systems. Their ignorance can be an obstacle, but one that's easily overcome. Be sure to contact your agent well before you buy your system.

When discussing insurance with your agent, explain things simply. Remember, he or she doesn't need all the details and probably wouldn't understand them anyway. Meet in person if at all possible so you can clear up any misunderstandings. It's much easier to interpret confusion in a voice or face. An email that summarizes the system and the location of the components may be helpful. Note that inverters and other equipment located inside a house may be covered by the household policy while a ground-mounted array requires a special rider.

Inform your agent that the solar electric system will be designed and installed by a professional according to the National Electrical Code and that all the equipment is UL-listed. Also let the agent know that the design and installation will be approved by the building department, if any, and that it will be inspected as well for compliance with local codes, even if you live in an area that doesn't require an electrical permit. (That's part of your interconnection agreement).

You may want to take pictures and diagrams of PV systems, but use them only if your agent seems confused. Keep the diagrams simple. This should allay any concerns your agent might have. If you need help, ask your installer to talk with your agent. He or she can explain how safe PV systems are, and why a huge liability insurance policy is not necessary. In my experience, insurance agents are basically just interested in what a system will cost to replace should it get damaged, so acquiring coverage has been pretty straightforward.

Whatever you do, always be forthright and honest with your agent. Let them know the full cost of the system, including installation. Not informing an agent of your intent to install a PV system could come back to haunt you. If you file a claim at later date, it is likely to be denied.

Buying a PV System

If a PV system seems like a good idea, I strongly recommend hiring a competent, bonded, trustworthy, and experienced professional installer. A local supplier/installer will supply all of the equipment, be certain that it's compatible, file for permits and interconnection agreements, and install the system. Your installer will test the system to be sure it is operating satisfactorily (don't pay in full until you are satisfied). They should be there to answer questions and to address problems, if any. Be sure to obtain written documentation of equipment warranties and any guarantee the installer makes regarding the system.

Consider hiring an installer who's bonded, especially if he or she is new to the business. A bond is sum of money set aside by contractors and held by a third party, known as a *surety company*. The bond provides financial recourse to homeowners and business owners should a contractor fails to meet his contractual obligations. In such instances, customers can file a claim for compensation from the bonding company. To stay bonded, a contractor must reimburse the company to cover payments made to dissatisfied customers. Residential contractors usually carry a minimum of $10,000 bond. Some AHJs require that contractors be bonded.

When contacting companies, ask if they are bonded and the amount of bond. Check it out. Be sure the installer has worker's compensation insurance to protect his or her employees when working on your site. Also ask for references, and be sure to call them. Visit installations, if possible. Talk to homeowners about their experiences and level of satisfaction with the installer, his crew, and the array. You may also want to check with the local office of the Better Business Bureau to check out an installer's rating. And, of course, get everything in writing. Sign a contract.

Whatever you do, don't pay for the entire installation up front. I ask for one-third upon signing a contract, so long as I can get to the installation within two to three weeks. That is the amount of time it takes to order and have equipment shipped to me. I ask for another third when I show up to begin the installation, and the last third when the building department and utility okay the system and it is up and running. The accompanying sidebar, "Questions to Ask Potential Installers," includes a list of questions you should ask when shopping for an competent installer.

You can also purchase equipment from a local or online supplier and install it yourself with or without their guidance, provided your AHJ allows property owners to perform electrical work. If you live in such an area, you can file for permits and install all equipment—PV arrays and equipment. Interestingly, however, in such instances, you cannot legally hire an unlicensed electrician to act as the electrical contractor. Nor can you employ any unlicensed electrician to work on the system.

Questions to Ask Potential Installers

1. How long have you been in the business? (The longer the better.)
2. How many systems have you installed? How many systems like mine have you installed? (The more systems the better.)
3. How will you size my system?
4. Do you provide recommendations to make my home more energy efficient first? (As stressed in the text, energy-efficiency measures reduce system size and can save you a fortune.)
5. Do you carry liability and worker's compensation insurance? Can I have the policy numbers and name of the insurance agents? (Liability insurance protects against damage to your property. Worker's compensation insurance protects you from injury claims by the installer's workers.)
6. Are you bonded?
7. What additional training have you undergone? When? Are you NABCEP certified? (Manufacturers often offer training on new equipment to keep installers up to date. NABCEP is a national certifying board that requires installers to pass a rigorous test and have a certain amount of experience.)
8. Will employees be working on the system? What training have they received? Will you be working with the crew or overseeing their work? If you're overseeing the work, how often will you check up on them? If I have problems with any of your workers, will you respond immediately? (Be sure that the owner of the company will be actively involved in your system or that he or she sends an experienced crew to your site.)
9. Are you a licensed electrician or will a licensed electrician be working on the crew? (State regulations on who can install a PV system vary. A licensed electrician may not be required, except to pull the permit, supervise the project, and make the final connection to your electrical panel.)
10. What brand modules and inverters will you use? Do you install UL-listed components? (To meet Code, all components must be UL-listed or listed by some other similar organization.)
11. Do you guarantee your work? For how long? What does your guarantee cover? How quickly will you respond if troubles emerge? (You want an installer who guarantees the installation for a reasonable time and who will promptly fix any problems that arise.)
12. Can I have a list of your last five projects with contact information? (Be sure to call references and talk with homeowners to see how well the installer performed and how easy he or she was to work with.)
13. What is the payment schedule?
14. Will you pull and pay for the permits?
15. Will you secure the interconnection agreement?
16. What's a realistic schedule? When can you obtain the equipment? When can you start work? How long will the whole project take?

You have to do the work yourself, except for the final grid connection, which must be performed by a licensed electrical contractor. The AHJ may also require the owner/installer to live in or occupy the building and may restrict him from selling or leasing the property for one year.

As a rule, I don't recommend installing a PV system yourself. The Code requirements are many, and complex. You really need to know your stuff just to secure a permit, let alone install the system safely. And, there are a lot of ways to mess up, especially with system grounding. Self-installs are appropriate for those that know a fair amount about electrical wiring and have lots of experience. You can gain some of this experience by attending several PV installation workshops.

Installing a PV system can also be risky. Working on electrical wiring is fraught with potential shock hazards. Even experienced installers get shocked once in a while or accidentally place themselves in very dangerous situations. Connecting to the electrical grid is a job best left for professionals, because working inside a main service panel can be extremely dangerous. Remember, a PV system is a huge investment and so are you, and you don't want to mess either up.

Another option is to work with an educational center like my organization, The Evergreen Institute, or the Midwest Renewable Energy Association. They're often looking for homeowners who are willing to sponsor a hands-on installation workshop. In my case, I design the system, order the equipment, and coordinate installation for a small upfront fee. I then advertise the workshop and make my money by charging a fee for each student. In cases such as mine, a qualified installer teaches the workshop and oversees all of the work (I am that person). Workshop registrants pay a fee to cover the cost of the instructor, the costs of advertising and setting up a workshop, and perhaps renting a portable toilet. The homeowner pays the costs of materials, and the workshop attendees provide much of the labor free.

Sponsoring a workshop requires advanced planning but could save a homeowner a substantial amount of money—I generally estimate savings of $1 per watt. For a 10-kW system, the savings could amount to $10,000.

Workshop installations can be fun and satisfying! You get a PV system installed on your home or business while providing a valuable opportunity for others to learn how to install systems. You do, however, have to be comfortable with a half dozen to a dozen more people working on your site for four or five days. If the workshop leader is competent and checks all of the attendees' work, you'll get a PV system installed right. Be sure to contact the nonprofit organization well in advance and be prepared to help organize and publicize the workshop. Also, be sure your insurance will cover volunteer workers on your site.

Parting Thoughts

When you started reading this book, you no doubt were already interested in PV systems. Perhaps you just wanted to determine if PV would be suitable for your home. Perhaps you were sure you wanted to install a PV system but didn't know how to proceed. I hope that you now have a clearer understanding of what is involved.

If you have come to realize that your dreams for a PV system were not realistic, be glad that you did not spend a lot of money on a PV system that would not have met your expectations. Efficiency measures are generally less expensive and more sustainable in the long run. If, however, you now have an informed conviction that a PV system is for you, you may want to contact a local installer or two. You can find them at local home shows, on the internet, and even in the Yellow pages.

You are now up to speed on photovoltaic systems. You know a great deal about solar energy, electricity, PV modules, types of PV systems, inverters, batteries, charge controllers, racks and mounting, and permits. You've gained good theoretical as well as practical information that puts you in a good position to move forward. I've given you mountains of advice on installation and helped you grapple with economic issues. Our work has ended, but yours is just beginning.

I wish you the best of luck and invite you to learn more at one of our workshops at The Evergreen Institute!

Summary of State Net Metering Policies

State	Subscriber limit (% of peak)	Power limit Res/Com(kW)	Monthly rollover	Annual compensation
Alabama	no limit	100	yes, can be indefinitely	varies
Alaska	1.5	25	yes, indefinitely	retail rate
Arizona	no limit	125% of load	yes, avoided cost at end of billing year	avoided cost
Arkansas	no limit	25/300	yes, until end of billing year	retail rate
California	5	1,000	yes, can be indefinitely	varies
Colorado	no limit	120% of load or 10/25*	yes, indefinitely	varies*
Connecticut	no limit	2,000	yes, avoided cost at end of billing year	retail rate
Delaware	5	25/500 or 2,000*	yes, indefinitely	retail rate
District of Columbia	no limit	1,000	yes, indefinitely	retail rate
Florida	no limit	2,000	yes, avoided cost at end of billing year	retail rate
Georgia	0.2	10/100	no	determined rate
Hawaii	none	50 or 100*	yes, until end of billing year	none
Idaho	0.1	25 or 25/100*	no	retail rate or avoided cost*
Illinois	1	40	yes, until end of billing year	retail rate
Indiana	1	1000	yes, indefinitely	retail rate
Iowa	no limit	500	yes, indefinitely	retail rate

State	Subscriber limit (% of peak)	Power limit Res/Com(kW)	Monthly rollover	Annual compensation
Kansas	1	25/200	yes, until end of billing year	retail rate
Kentucky	1	30	yes, indefinitely	retail rate
Louisiana	no limit	25/300	yes, indefinitely	avoided cost
Maine	no limit	100 or 660*	yes, until end of billing year	retail rate
Maryland	1500 MW	2,000	yes, until end of billing year	retail rate
Massachusetts**	6 peak demand 4 private 5 public	60, 1,000 or 2,000	varies	varies
Michigan	0.75	150	yes, indefinitely	partial retail rate
Minnesota	no limit	40	no	retail rate
Mississippi	N/A	N/A	N/A	N/A
Missouri	5	100	yes, until end of billing year	avoided cost
Montana	no limit	50	yes, until end of billing year	retail rate
Nebraska	1	25	yes, until end of billing year	avoided cost
Nevada	3	1,000	yes, indefinitely	retail rate
New Hampshire	1	100	yes, indefinitely	avoided cost
New Jersey	no limit	previous years consumption	yes, avoided cost at end of billing year	retail rate
New Mexico	no limit	80,000	if under US$50	avoided cost
New York	1 or 0.3 (wind)	10 to 2,000 or peak load	varies	avoided cost or retail rate
North Carolina	no limit	1000	yes, until summer billing season	retail rate
North Dakota	no limit	100	no	avoided cost
Ohio	no limit	no explicit limit	yes, until end of billing year	generation rate
Oklahoma	no limit	100 or 25,000/year	no	avoided cost, but utility not required to purchase
Oregon	0.5 or no limit*	10/25 or 25/2,000*	yes, until end of billing year*	varies

State	Subscriber limit (% of peak)	Power limit Res/Com(kW)	Monthly rollover	Annual compensation
Pennsylvania	no limit	50/3,000 or 5,000	yes, until end of billing year	"price-to-compare" (generation and transmission cost)
Rhode Island	2	1,650 for most, 2250 or 3500*	optional	slightly less than retail rate
South Carolina	0.2	20/100	yes, until summer billing season	time-of-rate use or less
South Dakota	N/A	N/A	N/A	N/A
Tennessee	N/A	N/A	N/A	N/A
Texas***	no limit	20 or 25	no	varies
Utah	varies*	25/2,000 or 10*	varies—credits expire annually with the March billing*	avoided cost or retail rate*
Vermont	15	250	yes, accumulated up to 12 months, rolling	retail rate
Virginia	1	10/500	yes, avoided cost option at end of billing year	retail rate
Washington	0.5	100	yes, until end of billing year	retail rate
West Virginia	0.1	25	yes, up to 12 months	retail rate
Wisconsin	no limit	20	no	retail rate for renewables, avoided cost for non-renewables
Wyoming	no limit	25	yes, avoided cost at end of billing year	retail rate

Note: Some additional minor variations not listed in this table may apply.
N/A = Not available.
Lost = Excess electricity credit or credit not claimed is granted to utility.
Retail rate = Final sale price of electricity.
Avoided cost = "Wholesale" price of electricity (cost to the utility).
* Depending on utility.
** Massachusetts distinguishes policies for different "classes" of systems.
*** Only available to customers of Austin Energy, CPS Energy, or Green Mountain Energy (Green Mountain Energy is not a utility but a retail electric provider; according to www.powertochoose.com).
Source: Freeing the Grid

Resource Guide

Books

Boxwell, M. *Solar Electricity Handbook, 2015 Edition: A Simple, Practical Guide to Solar Energy—Designing and Installing Solar PV Systems.* Greenstream Publishing, 2015. A book that covers the basics well, but leaves out some information installers need to know.

Chiras, Dan. *Green Home Improvement: 65 Projects that Will Cut Utility Bills, Protect Your Health, and Help the Environment.* RS Means, 2008. Contains numerous simple projects to make your home more energy efficient and reduce your energy bills.

Chiras, Dan. *The Homeowner's Guide to Renewable Energy.* New Society Publishers, 2/e, 2011. Contains information on residential wind energy and solar electric systems.

Dunlop, Jim. *Photovoltaic Systems.* American Technical Publishers, 2010. If, after finishing my book, you want to learn more, check out this book. It is a superb resource for future installers and other solar energy professionals.

Ewing, Rex and Doug Hargrave. *Got Solar? Go Solar: Get Renewable Energy to Power Your Grid-Tied Home.* PixyJack Press, 2005. Great little book for those interested in installing a grid-tied PV system.

Perlin, John. *From Space to Earth: The Story of Solar Electricity.* Aatec Publications, 1999. A detailed history of the development of PVs.

Schaeffer, John. *Solar Living Source Book.* New Society Publishers, 2014. Contains a lot of useful information on PV systems and PV products. I wrote the chapter on passive solar heating.

Solar Energy International. *Solar Electric Handbook: Photovoltaic Fundaments and Applications.* Pearson, 2011. A great book with wonderful illustrations for those interested in becoming an installer or individuals interested in installing their own PV systems.

Articles

Berman, Brad. "Fueling with Sunshine," *Home Power* 165, 42–46, 2015. Great article on powering electric vehicles with solar energy.

Calwell, Chris. "Efficient Home Lighting Choices," *Home Power* 165, 34–39, 2015. One of the best articles I've ever encountered on the efficiency of light bulbs.

Cohn, Lisa. "Safeguard Your RE Investment: Finding a Policy That Works for Your System," *Home Power* 128, 72–76, 2008. An insightful look at insurance for RE systems.

Dankoff, Windy. "Lightning Happens," *Home Power* 107, 60–64, 2005. Important information on protecting a renewable energy system from lightning.

Dankoff, Windy. "Top Ten Battery Blunders," *Home Power* 115, 54–60, 2006. Important reading for anyone installing a battery bank.

Davidson, Kelly. "Enervee: Scoring High Marks for Energy Efficiency," *Home Power* 161, 14, 2014. An interesting look at a new service that will help us determine the efficiency of appliances and electronic devices.

Donahue, Topher. "Solarscapes: A New Face for PV," *Home Power* 122, 36–40, 2007. Looks at an alternative way of mounting PVs to increase output and create shade.

Gocze, Tom. "Heat-Pump Water Heaters," *Home Power* 156, 58–64, 2013. Another must-read for anyone who wants a much less costly and more efficient alternative for solar hot water.

Gudgel, Bob. "Get Maximum Power from Your Solar Panels with MPPT," *Home Power* 109, 58–62, 2005. Excellent article on how maximum power point tracking works.

Guevara-Stone, Lauri and Ian Woofenden. "Choosing Your RE Installer," *Home Power* 127, 48–52, 2008. A good guide for hiring a professional installer for a PV system.

Hren, Rebekah. "Monitoring Batteryless Systems," *Home Power* 164, 30–36, 2015. Overview of ways to monitor PV systems.

Livingston, Phil. "First Steps in Renewable Energy," *Home Power* 118, 68–71, 2007. A good primer on home energy savings, vital to slashing the cost of a PV system.

Meyer, John and Joe Schwartz. "Battery Box Basics," *Home Power* 119, 50–55, 2007. Superb reference for individuals who want to build their own battery boxes.

Munro, Khanti. "Battery Monitoring," *Home Power* 128, 92–94, 2009. A general overview of battery-monitoring hardware.

Parker, Tehri. "Choosing the Right Site to Maximize your Solar Investment," *Home Power* 115, 30–33, 2006. A great description of how the Solar Pathfinder is used to assess solar resources and shading.

Perez, Richard. "Off-Grid Appliances: AC or DC?," *Home Power* 115, 52–53, 2006. A frank and useful discussion of DC appliances for anyone thinking of installing a DC only system or DC circuits to bypass the inverter for their off-grid home to save energy.

Russell, Scott. "Starting Smart: Calculating Your Energy Appetite," *Home Power* 102, 70–74, 2004. Great introduction to household load analysis (to determine household electrical demand).

Sanchez, Justine. "PV Energy Payback," *Home Power* 127, 32–36, 2008. Shows that PV systems produce as much energy as is required to make them payback the energy required to make them fairly quickly.

Sanchez, Justine. "PV Racks with Integrated Equipment Grounding," *Home Power* 166, 36–41, 2015. Important reading for anyone who wants to learn more about rack mounting options and grounding.

Sanchez, Justine. "Adding Battery Backup to Your PV System with AC-Coupling," *Home Power* 168, 38–43, 2015. Shows ways that grid-tied systems can be converted to grid-tied with battery backup.

Sanchez, Justine. "PV Modules…Updates and Trends," *Home Power* 169, 44–49, 2015. Good overview of improvements in modules.

Sanchez, Justine. "Railless PV Array Mounting," *Home Power* 171, 40–47, 2016. Well-illustrated piece on this increasingly popular system to mount modules on roofs.

Sanchez, Justine with Jsun Mills. "Adding Battery Backup to Your PV System with AC-Coupling," *Home Power* 168, 38–43, 2015. Very important reading for those who want to add batteries to their grid-tied systems.

Sagrillo, Mick. "Payback: The Wrong Question," *Windletter* 26(7), 1–3, 2007. Examines the topic of payback, especially why it is the wrong way to think about a renewable energy system.

Scheckel, Paul. "Efficiency Details for a Clean Energy Change," *Home Power* 121, 40–45, 2007. A quick guide to the ten most important energy efficiency measures.

Schwartz, Joe. "Solar Electric Tools of the Trade," *Home Power* 105, 22–26, 2004. Very well illustrated guide for do-it-yourselfers or an individual interested in becoming a PV installer.

Schwartz, Joe. "How to Install a Pole-Mounted Solar-Electric Array, Part 2," *Home Power* 109, 82–89, 2005. Extremely well-written and well-illustrated guide on installing a pole-mounted PV array.

Schwartz, Joe. "Finding the Phantoms," *Home Power* 117, 64–67, 2007. Excellent reading.

Schwartz, Joe and Zeke Yewdall. "Under Control: Charge Controllers for Whole-House Systems," *Home Power* 116, 80–84, 2007. Great overview of the various charge controllers on today's market.

Taylor, Jeremy. "Pump Up the Power: Getting More from Your Grid-Tied PV System," *Home Power* 127, 72–77, 2008. Good advice on ways to increase the efficiency of a PV system.

Tobe, Jeff. "The Right Fit: Hardware Solutions for PV Systems for Pitched Roofs," *Home Power* 161, 44–51, 2014. Excellent overview of various options to mount PV arrays on roofs.

Tobe, Jeff. "Managing Battery Charge Using Diversion Loads," *Home Power* 166, 52–58, 2015. Excellent overview of various options to mount PV arrays on roofs.

Truog, Jeremy. "Simple Steps to Save Through the Seasons," *Home Power* 115, 64–67, 2006. Excellent advice on saving energy in your home to reduce your PV system size and save money.

Weis, Carol and Christopher Freitas. "Battery System Maintenance and Repair," *Home Power* 161, 52–58, 2014. A must-read for anyone interested in a battery-based system.

Wilensky, Lena. "Grid-Tied System Performance Factors," *Home Power* 156, August and September, 2013. A must-read for installers.

Woodruff, Vaughan. "Solarize! Your Community," *Home Power* 171, 22–28, 2016.

Woofenden, Ian. "Battery Filling Systems of the Americas: Single-Point Watering System," *Home Power* 100, 82–84, 2004. This article is a must for those who would like to reduce battery maintenance.

Woofenden, Ian. "Managing Energy Use, One Load at a Time," *Home Power* 165, 16 and 18, 2015. An in-depth look at energy-efficiency.

Woofenden, Ian. "Off or On Grid? Getting Real," *Home Power* 128, 40–45, 2008. An in-depth look at the pros and cons of grid-connected and off-grid PV systems.

Woofenden, Ian and Chris LaForge. "Getting Started with Renewable Energy: Professional Load Analysis and Site Survey," *Home Power* 120, 44–47, 2007. A good article for those who are thinking about having a renewable energy system installed.

Yago, Jeff. "Portable PV," *Home Power* 168, 28–37, 2007. A good overview of the portable PV products available, from chargers to solar trailers.

Magazines

Backwoods Home Magazine. P.O. Box 712, Gold Beach, OR 97444. Tel: (541) 247–8900. Website: backwoodshome.com. Publishes articles on all aspects of self-reliant living, including renewable energy strategies such as solar.

Home Energy. 1250 Addison Street, Suite 211B, Berkeley, CA 94702. Tel: (510) 524–5405. Website: homeenergy.org. Great resource for those who want to learn more about ways to save energy in conventional home construction.

Home Power. P.O. Box 520, Ashland, OR 97520. Tel: (800) 707–6585. Website: homepower.com I have learned a ton from this publication; it's a goldmine of information. If you are interested in solar and other forms of renewable energy, I strongly encourage you to subscribe. HP publishes extremely valuable articles on renewable energy, including PVs, wind energy, microhydro, solar hot water, and passive solar heating and cooling. Magazine also contains important product reviews and ads for companies and profession installers. CDs containing back issues can be purchased from Home Power.

Mother Earth News. 1503 SW 42nd St., Topeka, KS 66609. Website: motherearthnews.com. One of my favorite magazines. Usually publishes a very useful article in each issue on some aspect of renewable energy.

Solar Today. ASES, 2400 Central Ave., Suite G-1, Boulder, CO 80301. Tel: (303) 443–3130. Website: ases.org/solar/. This magazine published by the American Solar Energy Society contains lots of good information on passive solar, solar thermal, photovoltaics, hydrogen, and other topics, but not much how-to information. Also lists names of engineers, builders, and installers and lists workshops and conferences

SolarPro. P.O. Box 68, Ashland, OR 97520. Website: solarprofessional.com. Superb resource for professional installers and the like.

Organizations

Center for Alternative Technology. Machynlleth, Powys SY20 9Az, UK. Tel: 01654 703409. Website: cat.org.uk. This educational group in the United Kingdom offers workshops on alternative energy, including wind, solar, and microhydroelectric.

Energy Efficiency and Renewable Energy Clearinghouse. P.O. Box 3048, Merrifield,

VA 22116. Tel: (800) 363–3732. Great source for a variety of useful information on energy efficiency.

Energy Efficient Building Association. 490 Concordia Ave., P.O. Box 22307, Eagen, MN 55122. Tel: (651) 268–7585. Website: eeba.org. Offers conferences, workshops, publications, and an online bookstore.

The Evergreen Institute, Center for Renewable Energy and Green Building. 3028 Pin Oak Road, Gerald, MO 63037. Tel: (720) 273–9556. Website: evergreeninstitute.org. Offers workshops on solar electricity, wind energy, passive solar heating and cooling, energy efficiency and much more through their educational center in Gerald, Missouri.

Solar Energy International. P.O. Box 715, Carbondale, CO 81623. Tel: (970) 963–8855. Website: solarenergy.org. Offers a wide range of workshops on solar energy, wind energy, and natural building.

Solar Living Institute. P.O. Box 836, Hopland, CA 95449. Tel: (707) 744–2017. Web site: solarliving.org. A nonprofit organization that offers frequent hands-on workshops on solar energy and many other topics. Be sure to tour their facility if you are in the "neighborhood."

U.S. Department of Energy and Environmental Protection Agency's ENERGY STAR program. Tel: (888) 782–7937. Website: energystar.gov.

PV Manufacturers

Advent solar: adventsolar.com
Canadian Solar, Inc.: csisolar.com
Day4Energy: day4energy.com
Kaneka: kaneka.com
Kyocera: kyocerasolar.com
Mitsubishi: mitsubishielectric.com/solar
Sanyo: sanyo.com
Schott: us.schott.com

Schuco: schuco-usa.com
Sharp: solar.sharpusa.com
SolarWorld: solarworld-usa.com
SunPower: sunpowercorp.com
Suntech Power: suntech-power.com
Sunwize: sunwize.com
Yingli: yinglisolar.com

Inverter Manufacturers

Beacon Power Corp: beaconpower.com
Fronius USA: fronius.com
Magnetek, Inc.: alternative-energies.com
Outback Power Systems: outbackpower.com

PV Powered Design, LLC: pvpowered.com
Sharp Electronics: sharp-usa.com/solar
SMA America, Inc.: sma-america.com
Xantrex Technology, Inc.: xantrex.com

Charge Controller Manufacturers

Apollo Solar: apollo-solar.net
Blue Sky Energy, Inc.: blueskyenergyinc.com
MidNite Solar, Inc.: midnitesolar.com
Morningstar Corp.: morningstarcorp.com

OutBack Power Systems: outbackpower.com
Xantrex Technology, Inc.: xantrex.com

Battery Manufacturers

Flooded Lead-acid

Deka/MK: eastpenndeka.com

EnerSys: hupsolarone.com

Exide Battery: hawkerpowersource.com

GB HUP: enersysmp.com

Power Battery: powerbattery.com

SunXtender: SunXtender.com

Surrette/Rolls Battery: surrett.com

Trojan Battery: trojan-battery.com

US Battery: usbattery.com

Sealed Batteries

Concorde Battery: concordebattery.com

Deka/MK: eastpenndeka.com

Discover Energy: discoverenergy.com

Exide: exide.com

Exide Sonnenschein: exide.com

FullRiver: fullriver.com

Hawker: hawkerpowersource.com

Power Battery: powerbattery.com

Trojan: trojan-battery.com

Rack (and Associated) Equipment Manufacturers

Direct Power and Water: directpower.com

EcoFasten Solar: ecofastensolar.com*

Ecolibrium Solar: ecolibriumsolar.com*

Ejot: ejot-usa.com

Hellerman Tyton: hellermantyton.com

Heyco: heyco.com

Ironridge: ironridge.com

Lumos Solar: lumossolar.com

Mounting Systems: usa.mounting-systems
.info

Mudge Fasteners: mudgefasteners.com

Pegasus Solar: pegasussolar.com*

PMC Industries: pmcind.com*

Quick Mount PV: quickmountpv.com*

PV Racking: pvracking.us

Roof Tech: roof-tech.com*

Renusol: renusolamerica.com

S-5!: S-5.com*

Silicon Energy: silicon-energy.com

Snake Tray: snaketray.com

SnapNrack: snapnrack.com*

Spice Solar: spicesolar.com*

Spider-Rax: spiderrax.com

UniRac: unirac.com

Zep Solar: zepsolar.com*

Zilla: zillarac.com*

*Manufacturers that offer railless PV roof mounts

Third-party Monitoring

Deck Monitoring: deckmonitoring.com

Draker: drakerenergy.com

eGauge: egauge.net

Locus Energy: locusenergy.com

Solar-log: solar-log.net

Battery Monitor Manufacturers

Bogart Engineering: bogartengineering
.com

OutBack Power Systems: outbackpower
.com

Xantrex: xantrex.com

Index

Page numbers in *italics* indicate diagrams.

About the Author

DAN CHIRAS is Director of the Center for Renewable Energy and Green building through which he teaches workshops on solar electricity, wind energy, passive solar design, natural building, and green building. He is also president of Sustainable Systems Design, Inc., a company that installs residential solar electric and wind energy systems and consults on passive solar design, residential renewable energy, and green building throughout North America. Dan is the author of 32 previous books, including *The Homeowner's Guide to Renewable Energy* and *Power from the Wind*.

A Note About the Publisher

NEW SOCIETY PUBLISHERS (www.newsociety.com), is an activist, employee-owned, solutions-oriented publisher focused on publishing books for a world of change. Our books offer tips, tools, and insights from leading experts in sustainable building, home-steading, climate change, environment, conscientious commerce, renewable energy, and more—positive solutions for troubled times.

The interior pages of our bound books are printed on Forest Stewardship Council®-registered acid-free paper that is 100% post-consumer recycled (100% old growth forest-free), processed chlorine-free, and printed with vegetable-based, low-VOC inks, with covers produced using FSC®-registered stock. New Society also works to reduce its carbon footprint, and purchases carbon offsets based on an annual audit to ensure a carbon neutral footprint. For further information, or to browse our full list of books and purchase securely, visit our website at: **www.newsociety.com**

New Society Publishers
ENVIRONMENTAL BENEFITS STATEMENT

For every 5,000 books printed, New Society saves the following resources:[1]

39	Trees
3,553	Pounds of Solid Waste
3,910	Gallons of Water
5,100	Kilowatt Hours of Electricity
6,460	Pounds of Greenhouse Gases
28	Pounds of HAPs, VOCs, and AOX Combined
10	Cubic Yards of Landfill Space

[1] Environmental benefits are calculated based on research done by the Environmental Defense Fund and other members of the Paper Task Force who study the environmental impacts of the paper industry.

MIX
Paper from responsible sources
FSC® C016245